Network Services
Investment Guide

Maximizing ROI in Uncertain Times

Network Services Investment Guide

Maximizing ROI in Uncertain Times

Mark Gaynor

Wiley Publishing, Inc.

Publisher: Robert Ipsen
Editor: Carol A. Long
Developmental Editor: Adaobi Obi Tulton
Editorial Manager: Kathryn A. Malm
Managing Editor: Pamela M. Hanley
New Media Editor: Brian Snapp
Text Design & Composition: Wiley Composition Services

This book is printed on acid-free paper. ∞

Published by Wiley Publishing, Inc., Indianapolis, Indiana

Published simultaneously in Canada

Limit of Liability/Disclaimer of Warranty: While the publisher and author have used their best efforts in preparing this book, they make no representations or warranties with respect to the accuracy or completeness of the contents of this book and specifically disclaim any implied warranties of merchantability or fitness for a particular purpose. No warranty may be created or extended by sales representatives or written sales materials. The advice and strategies contained herein may not be suitable for your situation. You should consult with a professional where appropriate. Neither the publisher nor author shall be liable for any loss of profit or any other commercial damages, including but not limited to special, incidental, consequential, or other damages.

For general information on our other products and services please contact our Customer Care Department within the United States at (800) 762-2974, outside the United States at (317) 572-3993 or fax (317) 572-4002.

Trademarks: Wiley, the Wiley Pubishing logo and related trade dress are trademarks or registered trademarks of Wiley Publishing, Inc., in the United States and other countries, and may not be used without written permission. All other trademarks are the property of their respective owners. Wiley Publishing, Inc., is not associated with any product or vendor mentioned in this book.

Wiley also publishes its books in a variety of electronic formats. Some content that appears in print may not be available in electronic books.

Library of Congress Cataloging-in-Publication Data:

Gaynor, Mark, 1956 –
Network services investment guide: maximizing ROI in uncertain times / Mark Gaynor.
   p. cm. — (Wiley Networking Council series)
Includes bibliographical references (p. 285) and index.
   ISBN 0-471-21475-2 (paper)
1. Telecommunication systems—Management. 2. Data transmission systems. 3. Integrated services digital networks. I. Title. II. Series.
   HE7631 .G26 2002
   384.3'3—dc21

                                                                                    2002014611

ISBN 0-471-21475-2

Printed in the United States of America

10 9 8 7 6 5 4 3 2 1

Books in Series

- *Understanding Policy-Based Networking*, Dave Kosiur
  ISBN: 0-471-38804-1

- *Delivering Internet Connections over Cable: Breaking the Access Barrier*,
  Mark E. Laubach, David J. Farber, Stephen D. Dukes
  ISBN: 0-471-38950-1

- *The NAT Handbook: Implementing and Managing Network Address
  Translation*, Bill Dutcher
  ISBN: 0-471-39089-5

- *WAN Survival Guide: Strategies for VPNs and Multiservice Networks*,
  Howard C. Berkowitz
  ISBN: 0471-38428-3

- *ISP Survival Guide: Strategies for Running a Competitive ISP*,
  Geoff Huston
  ISBN: 0-471-31499-4

- *Implementing IPsec: Making Security Work on VPN's, Intranets, and
  Extranets*, Elizabeth Kaufman, Andrew Newman
  ISBN: 0-471-34467-2

- *Internet Performance Survival Guide: QoS Strategies for Multiservice
  Networks*, Geoff Huston
  ISBN: 0-471-37808-9

- *ISP Liability Survival Guide: Strategies for Managing Copyright, Spam,
  Cache, and Privacy Regulations*, Tim Casey
  ISBN: 0-471-37748-1

- *VPN Applications Guide: Real Solutions for Enterprise Networks*,
  Dave McDysan
  ISBN: 0-471-37175-0

- *Converged Networks and Services: Internetworking IP and the PSTN*,
  Igor Faynberg, Hui-Lan Lu, Lawrence Gabuzda
  ISBN: 0-471-35644-1

Foreword

Networking Council Foreword

The Networking Council Series was created in 1998 within Wiley's Technology Publishing group to fill an important gap in networking literature. Many current technical books are long on details but short on understanding. They do not give the reader a sense of where, in the universe of practical and theoretical knowledge, the technology might be useful in a particular organization. The Networking Council Series is concerned more with how to think clearly about networking issues than with promoting the virtues of a particular technology—how to relate new information to the rest of what the reader knows and needs, so the reader can develop a customized strategy for vendor and product selection, outsourcing, and design.

In *Network Services Investment Guide: Maximizing ROI in Uncertain Times* by Mark Gaynor, you'll see the hallmarks of Networking Council books— examination of the advantages and disadvantages, strengths and weaknesses of market-ready technology, useful ways to think about options pragmatically, and direct links to business practices and needs. Disclosure of pertinent background issues needed to understand who supports a technology and how it was developed is another goal of all Networking Council books.

The Networking Council Series is aimed at satisfying the need for perspective in an evolving data and telecommunications world filled with hyperbole, speculation, and unearned optimism. In *Network Services Investment Guide*, you'll get clear information from experienced practitioners.

We hope you enjoy the read. Let us know what you think. Feel free to visit the Networking Council Web site at www.wiley.com/networkingcouncil.

Scott Bradner
Senior Technical Consultant, Harvard University

Vinton Cerf
Senior Vice President, WorldCom

Lyman Chapin
Chief Scientist, Founding Trustee of the Internet Society

Contents

Introduction

This book analyzes network-based services using a new framework that links market uncertainty to the choice between distributed management structure and centralized management structure. It explores communications technology and suggests how investors, managers, service providers, and consultants can make better decisions about technology. It also highlights how new technologies are adopted and evolve. This book introduces a new way to think about the management structure of network-based services such as email and voice mail.

In Part One, the critical link between market uncertainty and the choice between distributed and centralized management structure is explained, and examples that illustrate the power of this linkage are given. Part One explains why high market uncertainty favors distributed management, while centralized management does better when market uncertainty is low. Part Two contains two case studies of email and voice mail that demonstrate how this theory explains the history of these network services. Part Three applies ideas from Part One to several new technologies: Voice-over-IP (VoIP) wireless network infrastructure and Web applications and services. The chapter explores the predominant theme illustrated in examples from this book: Flexibility in choice between centralized and distributed management creates value for users and service providers. Last, I give advice to different types of readers of this book. For example,

I discuss how a venture capitalist should apply this theory to help decide when to invest in particular technologies.

This book presents an options framework to analyze the decisions about the architecture of network-based services and how these services should be managed, thereby giving managers a strategic advantage in today's uncertain world. The most important concept in this book is how to benefit from uncertainty by leveraging flexibility. Flexibility in uncertain markets creates value because it allows firms to optimize profitability by having a strategy that incorporates new information. Options illustrate the trade-offs between efficiency and flexibility — the greater the uncertainty, the greater the value of flexibility. Managers who know when flexibility has a greater value than efficiency (and other business advantages of centralized management structure) have a strategic advantage in today's uncertain and dynamic IT world.

A novel idea in this book is that choice in management structure is linked to market uncertainty. That is, the higher the market uncertainty, the greater the value of distributed management because it allows more experimentation. When market uncertainty is low, however, this benefit of easy experimentation evaporates, and the efficiencies of centralized management architecture become the dominating factor in this complex decision about management structure. Past approaches have based the choice of optimal management structure on any one of several different methodologies. Possible bases for the choice of management structure include whether it offers business or technical advantages and whether it allows wide latitude in the ability to create and implement new services. Some management structures offer the business advantage of knowing who the users are, as well as what they are doing. This increase in available information is a huge advantage to the network manager. A particular management structure may allow more efficient management, such as central distribution of software or more efficient use of resources. One management structure may have technical advantages over another in terms of how complex a feature is to design and implement. Other management structures promote innovation, but they may do so at the expense of business and technical advantages. Management structures that promote innovation, such as end-2-end structure, allow experimentation by many participants because of the relatively low entry barriers and freedom to try new ideas. Previous research, though, has failed to link market uncertainty to choice of management structure. The case studies in Part Two illustrate this link between market uncertainty and management architecture by explaining previous choices managers and vendors made in the circuit-switched public phone network and the packet-switched Internet.

This book demonstrates that high market uncertainty enhances the value of innovation. In low market uncertainty, the business and technical advantages of a particular management structure are more important than the value of innovation. When market uncertainty is high, management structures that are less efficient and have other business and technical disadvantages, yet allow easy experimentation, are justifiable because of the high value of innovation. On the other hand, if market uncertainty is low, experimentation is of diminished value because it is easy to meet the well-understood needs of the user. This indicates that the value of innovation is less than that of the potential advantages offered by management structures that are more efficient and have other technical advantages. Market uncertainty is linked to the management structure of a network service by showing how it affects the value of innovation relative to the business and technical advantages of a more efficient, but less flexible management structure. This unifies previous approaches into a more general theory, which will lead to choosing the most appropriate management structure for a network service in a given market condition. This general theory explains behavior in today's two most important, yet very different, networks: the Internet and the Public Switched Telephone Network (PSTN).

The telephone network and the Internet are very different: one is intelligent (telephone), and one is not (Internet) [1]. In many countries (including the United States), the telephone network works well in providing voice services to many, at prices most can afford. While the centralized structure of the smart telephone networks is successful, the innovation that has occurred in the "stupid" Internet towers above that within the intelligent telephone network. One reason for the tremendous innovation in the Internet is because of the end-2-end (see Chapter 3) argument that promotes experimentation by end users. Although the telephone and the Internet are fundamentally different, they have many similarities: Both give managers choices about how to manage the services provided over these very different networks. In addition, management can make choices about what to outsource and what to build/manage in-house, as well as how to design any internal systems. Finally, in both systems, market uncertainty affects the value of these choices to an organization's bottom line.

This book advances the end-2-end argument by illustrating what conditions are suited for end-2-end applications. At the outset, the end-2-end ideas focused on the technical advantages of simple networks with complex edges and ends. One reason that end-2-end architecture has great value is that users sometimes come up with better solutions, which may not be the result of greater creativity, but rather the result of more experiments. Furthermore, the value of user innovation depends on the market uncertainty: High market uncertainty implies tremendous gain when users can experiment, but low market uncertainty suggests that many users are unlikely to devise better services than those offered by a few large, centralized service providers. This book links the value of end-2-end application structure to market uncertainty, thereby helping managers and designers decide how to apply the end-2-end argument to maximize value.

Organization of the Book

This book is organized into three parts: general theory, case studies, and applying the theory to current technologies. The six chapters in Part One present an intuitive explanation of the theory linking market uncertainty to choice in management structure (this theory is the framework of this book and was developed for my Ph.D. thesis). Part Two contains two case studies that validate this theory. These cases are two of the most successful network-based services: basic/advanced voice services and email. These cases illustrate how market uncertainty was an important factor in determining what management structure users preferred in the past. Part Three applies theories in this book to several upcoming technologies, including Voice over IP (VoIP), wireless network infrastructure (802.11 and cellular), and Web applications and Web-based services. The book ends with the appendix, which presents a formal description of the theory.

Part One — General Intuitive Theory

Part One is an overview of important concepts, including definitions of network-based services and their management structure, the end-2-end argument, market uncertainty, and options. It is geared toward how managers and investors can apply the particular concepts to decisions they must make in today's IT world. Part One should be read by all readers.

Chapter 2 — Network-Based Services

Chapter 2 discusses network-based services: why these services have market uncertainty and what it means to experiment with network-based services. The framework for understanding different management structures is introduced by discussing the trade-offs between centralized and distributed management architecture and how users shift between different management structures for similar services. It illustrates network-based services in two important networks: the Public Switched Telephone Network (PSTN) and the Internet. It explains how changes in market uncertainty are a factor in deciding what management architecture will best meet users' needs. In the PSTN's long history, market uncertainty has cycled between high and low. The shorter Internet history has seen decreasing market uncertainty for mature services such as email, but high market uncertainty for new technologies such as VoIP, wireless services, Web applications, and Web services. The strengths of centralized management structure, such as efficiency, stability, and consistency, are contrasted to its weaknesses of being inflexible and hard to experiment with. Distributed management architecture has opposite attributes: It is flexible and easy to experiment with; however, it is not as good at efficiently using its resources and may be a less stable and consistent environment. The main point of this chapter is understanding how similar services can be managed with different management structures (centralized to distributed) and the advantages and disadvantages of each.

Chapter 3 – Internet End-2-End Argument

This chapter introduces the end-2-end design principle, as well as its history and meaning in today's Internet. Many (including me) believe that the end-2-end argument is one key reason the Internet has seen so much innovation — its basic idea is to keep the network simple, moving as much complexity as possible to the end points of the network. This end-2-end principle applied to applications suggests that the less information the internal network infrastructure knows about any application, the better. End-2-end argues for simple network infrastructure and complicated end devices. One example of an end-2-end protocol is the Transmission Control Protocol (TCP), which is a transport protocol used for the Web. TCP is end-2-end because only the two end points of the TCP connection know a connection exists. The history of the end-2-end argument is traced from its beginning, with ideas from cryptography and transaction processing, to newer applications of end-2-end principles in the modern Internet, such as business transactions where participants need authentication and authorization. The Internet has changed — trust is no longer assumed, which is leading to changes in the Internet's overall architecture. New ideas, such as Network Address Translators (NATs), break the end-2-end model in evil ways, such as breaking the rule of unique global addressing of end points. Other ideas, such as a transaction consisting of several end-2-end interactions (for example, authentication services), make sense in the untrusted world of today. The important concept in this chapter is to understand why the end-2-end argument has created so much value in the Internet and how to keep applying this principle to continue promoting innovation in the future.

Chapter 4 – Management Structure of Network-Based Services

This chapter provides a more detailed comparison of the attributes of centralized versus distributed management structure. Examples from email, voice, and information services are included. It presents a hierarchical framework that examines the choices management must make. From top-level decisions (such as outsourcing or not) to lower-layer decisions (such as choosing the structure of the network infrastructure when building and managing your own services), this multilevel approach is applied first to email — examining the choices between outsourcing and managing one's own email service. Next, different options for voice services — such as buying a Private Branch Exchange (PBX) or renting Centrex service from the telephone company — are explained in this hierarchical structure. Last, we look at providing information over the Internet, as well as the different choices management has — from outsourcing, to building a farm of servers, to utilizing a large mainframe to consolidate many information servers. The most important point of this chapter is to understand not only that there are many choices in how to manage network services, but also how these decisions affect the ability of users and others to innovate by experimenting with new services.

Chapter 5 – Intuitive View of Options Theory

The theory of options explains the value of choice in capital markets. This theory is an accepted tool in modern finance to value financial assets in uncertain markets and build investments to suit many different investors. Options theory illustrates the value of having flexibility, as well as how this value grows as uncertainty increases. This theory has been extended (real options theory) to value nonfinancial assets such as investments in research and development, building IT infrastructure, and valuing modularity when building complex systems. Without dwelling on the complex mathematics explaining the theory behind options and real options, this chapter provides an intuitive look at this important theory. The value of giving users choices in capital markets and how this value increases with greater market uncertainty will be discussed. Options theory illustrates the value of flexibility and thinking in terms of keeping one's options open. The most important concept in this chapter is to understand how the options framework explains why greater market uncertainty increases the value of having many choices.

Chapter 6 — Market Uncertainty

In Chapter 6, market uncertainty, as well as how to measure it and its effect on the value of experimentation, is explained in detail. Market uncertainty occurs when vendors, service providers, and even users don't know what will be successful. Historically, market uncertainty for Information Technology has been high, with wrong predictions being the rule. Consider Asynchronous Transfer Mode (ATM) and Integrated Service Digital Network (ISDN): two technologies that were predicted to become winners, but that never became successful. After further defining market uncertainty, this chapter explores different methodologies to gauge its level. These include mythologies used by others, as well as new methods explored in research for my Ph.D. thesis. The chapter ends by linking the level of market uncertainty to the value of experimentation: the greater the market uncertainty, the larger the gain from experimentation. Chapter 6 illustrates how low market uncertainty translates to little value of experimentation because all the experiments are close to average. High market uncertainty, though, spreads out the results of many experiments — some results are terrible, being way below the mean, while other results are fantastic, having values far greater than average. The critical point of this chapter is to understand how to gauge the level of market uncertainty and why it is linked to the value of experimentation (the higher the market uncertainty, the greater the value of experimentation).

Chapter 7 — Theories about How to Manage Network-Based Services

Part One ends with a short chapter that puts everything together and presents a theory explaining how market uncertainty is an important factor in deciding how to manage network-based services. Chapter 7 first scopes services to which the theory applies and then illustrates the trade-offs between the business and technical advantages of centralized management structure and the flexibility and easy experimentation of distributed management architecture. This chapter weaves together the previous chapters in Part One into an intuitive and easy-to-understand theory. The most important point in this chapter is to understand why market uncertainty is a factor that managers and investors must consider in order to maximize value when choosing the best management structure for the current environment and how flexibility in this management structure creates value in dynamic environments.

Part Two — The Case Studies

Part Two consists of cases studies of two popular network-based services: email and voice. Each case starts with a discussion of the service's history, which illustrates concordance with the theories from Part One. Next, shifts in management structure are shown to occur in coordinating to changes in market uncertainty. These cases illustrate how important market uncertainty is as a force shaping the management architecture that works best. These two cases demonstrate the strength of the theory because it explains the evolution of two very different network services, in two very different types of networks (PSTN compared to the Internet). Each service, in each network, has evolved by giving users choices in how to manage these services.

Chapter 8 — Email

Email is one of those services that I find hard to believe I ever lived without. It has become a mainstream and ubiquitous way of communication. Email's evolution fits well within the framework based on the theory from Part One: It started out with a distributed management architecture and evolved such that users have choices between email systems with distributed or centralized management structures. At birth, email consisted of one user sending another user a file over a file transfer protocol such as FTP. Today, email service providers, such as Hotmail, that provide email with a centralized management structure have the fastest growth. The current environment allows choice in how to manage your email services, from the most distributed to the most centralized. This case examines the shift of users from distributed to more centralized email architecture in the mid-90s — at the same time that market uncertainty was low by several measures. The evidence argues that market uncertainty is the main factor causing this shift in management structure from a distributed structure, ISP-based email to a more centralized, Web-based email architecture. The important concept in this chapter is that market uncertainty can cause shifts in management structure (as seen in the shift of users to a centralized email structure).

Chapter 9 — Voice Services

Voice has a long history, but for most of it, market uncertainty was very low because of regulation. Until recently, users had only one choice for voice service or equipment — AT&T was the only option by law. The regulatory environment relaxed starting in 1968 with the *Carterfone Decision*, which gave users choices among equipment and, later, services. By the 1970s, users had many options for voice services: PBXs from several vendors and Centrex service from the phone company. This case study looks at where most of the innovation has occurred with voice services; the distributed PBX is marked as a source of much of this innovation. It is illustrated that users preferred PBXs (a distributed architecture) when market uncertainty was high but migrated to Centrex (a more centralized way to provide PBX-like services) as market uncertainty decreased. This case examines basic voice services, such as caller ID or call forwarding, and more advanced voice features, such as voice mail and Automatic Call Distribution systems (ACDs), illustrating the similarity of how these features evolved. First, they were implemented with distributed architecture; after proven successful, these services migrated into the network via Centrex services. The important points of this chapter are to understand that users had many options to manage voice services and that market uncertainty was a main factor in making these complex decisions.

Part Three — Applying Theory to Current Technology

Part Three applies theories from Part One to current technologies, including Web-based applications and services, Voice over IP, and wireless network infrastructure. It also gives advice to the different types of readers of this book, which include investors and managers in many different industries. This part tries to do the impossible: predict how these important technologies will evolve. Part Three notes a common thread in what is becoming successful: Protocols that allow flexibility in the choice of management structure appear more successful than those that restrict the choices managers have. From Voice-over IP to wireless infrastructure and Web-based applications and services, the protocols being adopted by vendors and users allow a continuum of management structure from distributed to centralized. This part emphasizes the power of the theories from Part One because of their applicability to such different technologies.

Chapter 10 — Voice-over IP (VoIP)

The convergence of voice and data is now unfolding, with the outcome very different from what telephone companies envisioned: Instead of the Public Switched Telephone Network (PSTN) becoming the transport for data, the Internet is becoming the transport for voice. VoIP is the technology of sending voice over the Internet — it's becoming popular as the technologies of the Internet mature and meet the rigid Quality of Service (QoS) required when sending voice over packet networks. As with most new technologies, there are different proposals for how to best provide voice over the Internet. One idea from the telephone companies called megaco/H.248 is similar to the current design of the PSTN. Megaco/H.248 forces users into a model of centralized management — something very familiar to those championing this technology. This is both a good and a bad thing: It's good in that we know the architecture will work, but it's bad because it doesn't utilize forward thinking. Could a different paradigm create architecture of more value to both users and service providers? Session Initiation Protocol (SIP) is a better model. It is flexible in what it allows. Users can choose either a centralized or a distributed management structure — it is their choice. Again, the important point of this chapter is that this flexibility creates value because it better meets the needs of more users.

Chapter 11 — Coexistence of 802.11 and 3G Cellular: Leaping the Garden Wall

Wireless devices have always been popular. From remote controls to wireless phones, being untethered is what users want and need. Technology has recently made it possible to have wireless Internet connectivity. Many users connect to a local LAN with 802.11 technology because it has become inexpensive and works well. Big carriers are spending billions on the next-generation cellular networks (3G, for now). Are these two wireless technologies substitutes, such as coffee and tea where you want one or the other but not both, or are they complements, such as coffee with milk, that when combined make each more valuable? The big service providers hope users don't need 802.11 technology once their 3G networks are rolled out, but many users believe differently. This chapter argues for a complementary point of view and illustrates how both cellular and 802.11 users, vendors, and service providers will capture the most value if they share a friendly coexistence. It demonstrates how the strengths of each technology match the weaknesses of the other technology, demonstrating that users will be better served if both technologies work together. The important point of this chapter is that giving users interoperability among complementary technologies creates the most value for the most users.

Chapter 12 — Web-Based Applications and Services

Web-based applications have changed many things for many people. I bank online, get information about my students online, make travel reservations, and perform many other tasks online. Web-based applications have become the standard way to obtain information and request services for good reason — they work well for many users. Web-based applications have properties of both distributed and centralized management structure. It's easy to experiment with Web-based applications, but once successful, they can be managed with a centralized architecture (for example, Hotmail). Having this flexibility of management structure is extremely powerful — it's the best of both worlds: easy experimentation and efficient use of resources.

The new breed of Web-based services based on data encoded with Extensible Markup Language (XML) and delivered in a Simple Object Access Protocol (SOAP) envelope is being embraced by all the major vendors including Microsoft, Sun, and IBM. Is this a new idea or just the next phase of Electronic Data Interchange (EDI), or both? Web-based services have many things going for them — vendors, service providers, and even users are starting to believe in their value. Vendors don't agree on the implementation details, but they do agree on the big picture. Web-based services have attributes similar to Web-based applications: They are easy to develop and experiment with, but these Web services can also take advantage of the technical and business advantages of centralized management structure. Again, we see great flexibility in how Web services can be managed — from distributed to centralized. This flexibility, allowing a continuum of management structure, is one key factor for capturing the most value from Web applications and services. The important point of this chapter is to understand how Web applications and services allow distributed management structure, promoting experimentation and innovation, as well as how these services benefit from the business and technical advantages of centralized management.

Chapter 13 – Conclusion and Advice to Readers

The final chapter explains the generality of the theories in Part One to different types of networks with different types of services. In both the Internet and PSTN, services have evolved with both centralized and distributed management structures. A common thread is found with all the technologies we looked at (email, voice services, Web applications and services, VoIP, and wireless network infrastructure): The successful services and protocols allow a continuum of management structure from distributed to centralized. The chapter and book end with an attempt to offer advice to the different groups of readers expected to read this book: investors, managers in many different industries, and consultants. Each group has different goals and should therefore use the theories in this book differently. Investors and venture capitalists need to maximize their return on investment by knowing when to invest in particular technologies. Managers will have different goals, depending on what industry they are in because vendors, service providers, and service users have very different needs. This chapter focuses on how each of these diverse groups should think about this book. Naturally, different readers will apply the information in this book in varying ways.

Appendix — Formal Theory and Model

The appendix is a more formal model than that presented in Chapter 7. The language is more academic, and the mathematical details are provided for the interested reader. This appendix is composed of two parts: formal assumptions and theory. The mathematical model is based on real options and has strong theoretical foundations. The model is accessible to anybody with a basic understanding of mathematics. Even though the equations might look intimidating, it is mostly the notation that is complex, not the mathematical techniques. The model is presented in two stages: first the simple model, followed by a more complex model that incorporates how vendors, service providers, and users learn as technology matures. The important point in the appendix is the analytical framework underlying the theories presented in Part One.

How to Read This Book

As the author, I recommend reading this book cover-to-cover, but I recognize this might not be the right approach for all readers. Part One of this book is the most important because it builds the framework behind linking market uncertainty and the choice of management structure. Without this framework, the other two parts will be difficult to understand. Part Two presents two case studies (email and voice mail) and is optional reading. If you are interested in how technology evolves and is adopted by users and/or the history of these services, the case studies will be interesting. Part Three illustrates how to apply theories from Part One to current decisions concerning choice in management structure. Most readers will be interested in this part because it discusses technology topics of interest to most academics and practitioners of management and information science.

The next chapter introduces network-based services and their management structure. It explores the strengths and weakness of both distributed and centralized management architectures. It links the ability of users to experiment with management structure, and it shows that although distributed management structure promotes easy experimentation, it may have business and technical disadvantages when compared to centralized management.

Background, Framework, and Theory

Network-Based Services

Understanding network-based services is a critical success factor for any manager in today's world. Managers have a wide variety of choices in deciding to outsource a network-based service or build the infrastructure to offer it in-house. Furthermore, whether outsourcing or building in-house, managers must make choices about the management architecture of such services. For example, consider voice services and the many ways to provide them within the office. You might go to Radio Shack and just buy a commodity PBX, install it, and manage it; this is a distributed solution. On the other hand, you might subscribe to Centrex, a service from the local telephone company that mirrors the services of a PBX. This chapter will help you develop another way to think about the structure of network-based services in the context of meeting users' needs when these needs are hard to predict.

There is a need for a new way to think about the infrastructure of network-based services. Past approaches about how to decide the optimal management structure are limited because they focus on either the business or technical advantages or whether a management structure offers wide latitude in the ability to create and implement new services. The methodology used in this book balances the business and technical advantages with the value of innovation by linking market uncertainty to the value of experimentation.

This chapter discusses what a network-based service (NBS) is and how it can have market uncertainty. It explores experimentation in the context of network-based services and how it helps to meet market needs. It examines different ways to manage a network-based service, illustrating how network services can have either distributed or centralized management architecture. It introduces the connection between market uncertainty and choice of management style. High market uncertainty is linked to the success of distributed management, while low market uncertainty implies the likely success of a centralized structure.

Defining a Network-Based Service and Its Market Uncertainty

Network-based services are hard to define precisely because of the diversity of networks and users. Some network-based services familiar to users of the Public Switched Telephone Network (PSTN) are basic voice services, such as transferring a call and speed dialing, and advanced services, such as voice mail and Automatic Call Distribution (ACD). Network-based services provided over the Internet include email and the name-to-address translation that the Domain Name Server (DNS) provides, along with a plethora of Web-based applications and services. Some network services such as email and voice mail are visible to the network users. Others, such as circuit setup in the PSTN and hop-by-hop packet routing within the Internet, provide a basic infrastructure that is invisible to most users. The operative definition in this book is that a network service is any service provided within or over a network.

When a service becomes available on the market, it may or may not be possible to accurately predict how well its features will meet market demands. This happens because of market uncertainty (MU), which is the ability to predict the market for a particular feature set of a service. There are many examples illustrating high market uncertainty with network-based services, such as the prediction that video phones would be popular before 2000, AT&T's predictions about the small potential size of the wireless phone market[1], or the predicted success of ATM to the desktop. Market uncertainty may also be low, such as in the case of basic email and basic voice features that are known to meet user needs well. Market uncertainty

---

[1] Presented in class about network architecture by Scott Bradner 2/09/01.

is this ability to understand and predict what users will embrace, and what they will not.

Previous research from Clark [1] shows that when new technology is first introduced, users' expectations evolve along with the technology. Clark noted that when the first automobiles were built, users viewed them in the context of a horse-drawn carriage (hence the name "horseless carriage"). Only later, as users began to understand the range of possibilities, did attributes such as reliability, comfort, and safety become important. A similar phenomenon is occurring with the Internet and the Web. The diversity of Web-based applications is beyond what pundits ever imagined. Nobody predicted in the early 90s what the Web is today or its impact on society. The combination of new technology and users' perceptions of their evolving needs creates market uncertainty.

One important question about market uncertainty is to whom does this market uncertainty apply. Is it the user of the service or the manager of services within a company? The answer may be very different depending on whether the usage is within a corporation or private. For example, consider email; for end users at home buying this service, the market uncertainty is relative to the service provider's ability to meet the needs of the email user. The corporate situation, though, is more complex. Typically, at a company, a network manager is responsible for providing telecommunication services. This manager pays the service provider for service and may bill different departments for services provided by the networking group. The market uncertainty that this manager faces may have two components. First, company policies may dictate what the network group can do. Concerns about security or preferences of upper management may constrain the choice of vendor and type of services the manager is allowed to provide to company employees. Second, if company policies allow some discretion, the manager is responsible for keeping the users happy. This shows that market uncertainty is complex because of its many dimensions.

Looking at how Harvard University provides services to its users illustrates this better. Harvard is most concerned with meeting the needs of its academic and management populations. Management at Harvard has two basic choices for email: buy "raw" IP service and use it to provide mail servers owned and managed internally (the current solution), or outsource the service. By providing email in this distributed fashion, Harvard has control over its email system. Because the uncertainty of basic email is low and Harvard's record for managing its email system is not perfect, I believe (by thinking along the lines of this book's argument) that in the current email environment the needs of Harvard users can be met by outsourcing email. Harvard also provides telephone services to its users and has two choices: the distributed solution, where the university owns and manages its PBXs, or the centralized model of Centrex. With normal voice over the PSTN, Harvard uses the Centrex service. Market uncertainty is low in this area, and the needs of the users are well met. Harvard, however, is now experimenting with Voice-over IP, a new area with much market uncertainty about what features customers want. Harvard feels the needs of its users are better met in this particular case with the distributed solution of owning an IP-PBX. Leo Donnelly, the manager of this project at Harvard, knows that he does not understand what his users want from Voice-over IP, so he is allowing experimentation. This example shows that for Harvard, market uncertainty is focused on the end user of the service because of the diverse group of users with very different needs. If the network managers at Harvard do not keep their users happy, others who can provide good services will replace them.

Today Web-based applications are the rage, but how does one decide which Web-based application is best for a particular business need? Web-based applications are very diverse, from traditional Internet services, such as email, to a new breed of services enabled by the interactive nature of modern Web pages. From taxes to maps to banking, Web-based applications are trying to find the right business models. Some useful services such as travel directions and maps (MapQuest) are popular and provide a tremendous service, as Scott Bradner pointed out in a *Network World* column [2]. It is unfortunate that these companies do not yet have a business model that can generate revenue for well-received services. At present, nobody knows what services will succeed from a business standpoint by providing sufficient value to those supplying the service. This book helps both the manager and investor by illustrating how market uncertainty affects the value of Web-based applications.

Companies such as Microsoft, with its .NET, and Sun, with J2EE, are betting that users want to use Web-based services. They believe that a big market exists for applications built with Web-based services. For now the market uncertainty in this area is huge. Not only are the services that business wants unknown, but the best architecture with which to build these services is unclear. Microsoft believes its .NET scheme is best, but other major players such as Sun and IBM believe in a pure Java environment. While it is almost certain that Web-based services will be adopted by many businesses, what the services are and how they will work are currently the source of much industry speculation, indicating high market uncertainty.

There is much uncertainty about the adoption of these network-based services by businesses. Will companies build their own services, or will they use service providers? Furthermore, what will the management structure of these services be? Outsourcing will allow businesses to focus on their core companies, while building their own services will allow more flexibility in meeting uncertain user needs. The jury is out; we now need to wait to see what happens.

In today's IT environment, managers and investors who understand why particular network-based services have become successful over other services are in a better position to decide what type of network infrastructure is best, given the degree of market uncertainty. Some network-based services, such as basic voice and email, have changed how we live and work and have created vast wealth. Other services, such as videophones, have never emerged as predicted. This way of thinking with an option point of view helps manage the uncertainty in the dot-com and telecom sectors, capturing the most value from these services by minimizing the risk associated with high market uncertainty.

Network Service Experiments

When service providers do not understand the needs of their users, service providers that experiment and give users many choices have a higher probability of succeeding in the uncertain market. Each experiment represents a feature that has unknown value to the user. The value might be high, as in the case of providing voice mail features within a PBX, or low, such as features that multiplex data and voice in PBXs. Each experiment represents an attempt to meet the needs of the user.

The value of network service experiments is determined by market selection. The accuracy of any estimate of service value depends on the level of market uncertainty, which is a measure of variance of the randomness in the market. This value is the relative success of the service experiment in contrast to the other options the customer has. The reflection of this value is the ratio of users that one particular service has over another. This is not a question of the kind of architecture that a specific company will use. A company may always use a central model even when this theory says that will not be the winning strategy. The market determines the value of the service by allowing users to pick what meets their needs best. Vendors can't predict what users want, and users can't tell vendors what will satisfy them, but users know what they like after they see it.

How difficult each experiment is to run and who can experiment are important attributes of management structure. Was permission required of the network owner/managers? How technically difficult is the typical experiment? What does it cost? How questions like this are answered affects the amount of experimentation that occurs and thus the value of the most successful of them. A greater number of experiments indicate a greater expected value. In general, services with distributed management structures allow more experimentation while centralized managed services imply less, as explained further in this chapter.

While general services are predictable in many instances, the particular architecture, feature set, and implementation that are widely adopted are often not predictable. Email is a good example of this; the demand was clear, but it took several generations of competing service offerings to converge to the Internet standards-based solution. In the 80s, X.400 became the anointed email standard; it met everybody's needs, according to its designers, and was championed by most governments and vendors. It seemed that it could not fail because it had options for every need, but it did fail. By the early 90s, the adoption of Internet email seemed to indicate that it could become a serious threat to X.400, yet the majority of industry pundits adamantly still believed in the OSI's X.400. These pundits asked how X.400 could fail; it was feature-rich compared to Internet email, which could not send attachments or demand proof of mail delivery. The U.S. government even mandated that X.400 be addressed in all responses to RFPs with the U.S. Government OSI Profile (GOSIP) directive. The success of Internet email over X.400 illustrates that even when demand for services such as email is understood, the particulars of the service are not. This indicates that current predictions may be inaccurate for services and feature sets of new technologies such as Voice-over IP, wireless, and general network-based services as the convergence of voice and data becomes a reality.

One important part about experimentation with network-based services is the selection process by which the services become successful, which is similar to how the fittest organisms are selected within bio-environments. Users select the "best" service — after they see the choices, similar to how bio-environments pick the best choices from many genetic mutations. Value determination of network service experiments is made by market selection by the users, similar to how bio-environments value a genetic mutation based on how well it survives and passes on the altered gene.

Value is related to user adoption, which is related to meeting user's needs. There are many examples of both successful and unsuccessful network-based services. In the telephone network, caller ID and personal voice mail have been well received; however, video calling and ISDN have not. There are similar examples in the Internet, with email and Web-based applications such as the CNN or the *Wall Street Journal* site being incredibly successful, while other Web-based applications, such as food delivery, have been a dismal failure. Even multicast has not been adopted by most Internet users. The market uncertainty is what makes it hard to predict what services with what feature sets will best meet user needs.

A very important question is who can experiment with a network-based service? In the telephone network, it is impossible to add new services within the network. For example, in the PSTN, consider the *## services [3], such as retrieving the number of the last incoming call. Users cannot change or add new *## services; only the network telephone company can experiment with them. The Internet is fundamentally different because it promotes users' ability to create their own services and provide them to others. It is easy for anybody to add new services. The network services provided by many Internet service providers are basic IP connectivity and email. The PSTN does allow some experimentation because PBX vendors not related to the telephone company can experiment. In addition, within the Internet, proposed services, such as Voice-over IP with the megaco protocol, depend on a centralized network server. While hard to categorize, the Internet in general promotes easy experimentation by all, while the telephone network does not.

The ease of experimentation is also important. How complex must experiments be, and what resources are required to perform them? Does network infrastructure need to change to support the new service or feature? To change a telephone *## service, the Central Office switch needs modification, but when the Web was created there was no change to the Internet infrastructure. Some Web technologies such as Java empower edge innovation because it pushes the intelligences of Web applications to the network's edge. Shifting more processing to the users' desktops via local execution of Java applets encourages end users to experiment, as the end-2-end argument discussed in Chapter 3 explains. When experimentation is easy, it happens more. When more people can experiment, it happens more. Easy and inexpensive experimentation promotes more experimentation — a good thing when users' needs are a mystery.

As discussed in Chapter 6, when the variance of a set of random experiments is high, the results of the experiments are widely dispersed from very good to very bad. It is the combination of easy experimentation, the ability of end users to experiment, and high market uncertainty that can work magic. These are exactly the conditions under which the Web was created. When market uncertainty is low, experimentation is of little value because anybody can meet user needs. The important point in low market uncertainty is the ability to provide a known good service to users at the best price for a given set of features and performance. It is market uncertainty that determines if experimentation makes sense from the point of view of meeting user needs.

Introduction to the Management Structure of NBS

Different management structures promote different advantages. Sometimes the ability to experiment is the most important attribute of a management structure; this indicates that a distributed structure will work best. In cases when market uncertainty is low, however, business and technical reasons justify a more centralized management structure that is efficient in its use of resources and allows tracking of its users. Both centralized and distributed management structures have advantages and disadvantages; the choice of the most appropriate structure depends on the situation. This is similar to the theory of contingency management discovered by Lawrence and Lorsch [4], the main contribution of which is this: The best management structure depends on the organization and the environment within which it functions. This theory is a contingency theory of service management: The best management structure for a network-based service depends on the amount of market uncertainty in the marketplace. The management structure of a service is not to be confused with the management structure of a company. The management structure of a company explains the hierarchy of authority and is not related to the choice of structure the company imposes on the services it provides. Small ISPs with central management offer more distributed email service than companies that have a more functional management structure such as Microsoft, yet offer a more centralized type of email service. The management structure of a PBX vendor is unrelated to the fact that PBXs are a distributed solution. The choice of management structure does matter, and this choice is related to market uncertainty.

Figure 2.1 shows an example of two network services with different management structures. First, in (a), is the 800-number service provided by the phone company. This service allows a user to buy an 800 number from a service provider where the fee for a call to the number is billed to the user being called (not the calling party, which is normal). This is an example of a service with central management. The phone network does not know how to route 800 numbers, but it does knows how to route *normal* phone numbers with an area code because the prefix locates the phone number. To provide this service, a centralized database holds 800 numbers and their corresponding *normal* phone numbers. When the network sees the 800 number, it sends a message across the SS7 network to query the 800 database; the result is a phone number the network knows how to route. This is an example of a centralized structure used to provide a name resolution protocol to translate the 800 number into an area code and telephone number. This management structure is centralized because a central authority manages both the physical 800-number servers and the data on the server. This is not the only way to structure such a protocol, as seen in (b), which illustrates the distributed structure of the Domain Name Server service in the Internet. Its management structure is distributed because both the DNS servers and the data on the DNS server are managed by the local organization. For example, if a host in the bu.edu domain seeks the IP address for a host in some other domain under the .com prefix, then the DNS sends a series of messages, eventually reaching a DNS server managed within the domain of the requested address.

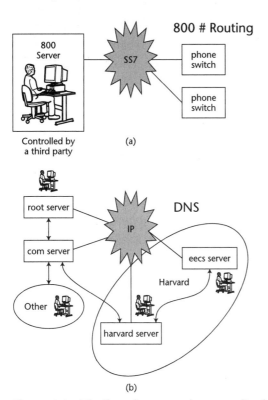

Figure 2.1 Distributed compared to centralized management structure of network-based services.

Many services with centralized management, such as the 800-number lookup, provide a carrier revenue opportunity. This is unlike the DNS service where a revenue model does not make business sense. Centralized management lends itself to billing for services; for example, the centralized architecture of the 800-number lookup makes it easy to track use because 800-number lookup requests go to a few centralized servers that are managed by a central authority. Logging their use is easy because the information about who is doing what is centralized. The distributed nature of the DNS architecture would make billing much harder. Most low-level DNS servers responsible for local IP addresses are managed by the local organization. The information about usage is very distributed, which adds complexity to billing. DNS caching creates more billing headaches because it distributes usage information even more. Centrally managed services, such as 800-number lookup, lend themselves to easy billing of the service, but services with a distributed management structure, such as DNS, make it difficult to know what users are doing, which creates billing problems.

The DNS is a service with a distributed management structure that is justifiable even in the absence of market uncertainty. Because organizations value control over the data that defines their internal address space, distributed management of DNS is desirable for many organizations. DNS is a distributed database used to resolve a name to IP address, a service similar to the 800 lookup service, but with a different management structure. This underscores that services with similar functions may be architecturally dissimilar.

The different ways that email can be stored and routed are a good illustration of how both centralized and distributed management structures can provide similar services. Chapter 8 will show that initially Internet email was completely end-2-end. Later, a partially centralized and partially distributed architecture for Internet email used POP or IMAP[2] to fetch email messages from a remote server; the retrieved emails were then managed by the local system. New Web-based email systems forgo the POP/IMAP interface and access a central database of email via a Web interface. This (central) management of messages occurs on the email server, not on the local system.

Different networks, such as the PSTN and the Internet, have different infrastructures that tend to promote different styles of management, but each network allows some services to be provided in either a centralized or a distributed structure. For example, as discussed in Chapter 9, there are different ways to provide basic and advanced voice services in the PSTN.

[2] IMAP, which came after POP, is more centralized because it allows the remote server user to view the message's subject while the message is still managed by the IMAP server.

Users might choose a PBX at the user's site (distributed) or a Centrex service provided within the Central Office (CO) of the local phone company (centralized). Internet email may have either a distributed or a centralized management structure. A user running Sendmail on a desktop or using POP to communicate to an email server is a more distributed situation than a user employing a Web-based email system such as Hotmail or Yahoo! mail. Web-based email is more centralized for two reasons. First, its servers are managed by a centralized organization. Second, the central server, not the user desktop, manages email messages. Again, these examples illustrate how networks with a centralized infrastructure are able to provide services managed with a distributed architecture and how networks with a distributed infrastructure, such as the Internet, are able to provide services with a centralized structure.

The scope of management relative to the organizational affiliation of the users is one attribute of a management structure that helps determine how centralized the structure is. In the preceding 800-number example, a central authority manages the server for all users, regardless of where the user (or his or her organization) is or where the called 800 number resides. In contrast, the likely manager of the DNS server responsible for the low-level IP address-to-name resolution is the organization that manages the host of the requested IP address. In the first example, the manager of the server has nothing to do with either user. In the DNS example, however, the manager of the first server queried is in the domain of the requesting host, while the management of the server that finally responds with the requested data is in the domain of the host with the destination address. In email, the situation is similar; ISP email is managed by the ISP. Its users are paying members of the ISP community. Web-based email, though, is managed by a central authority (for example, the management of Hotmail); its users don't belong to any common ISP. The important point is this: There are no distributed organizations managing the service. Scope of management is one way to classify the management structure of a network-based service.

Another attribute of a management structure is the style in which it manages its data. In the case of the 800 service, a central manager is responsible for the data regardless of where the caller or called party resides. In contrast to this is the management of the DNS (Figure 2.1(b)), where data is most likely to be managed in the same domain as the targeted host. ISP email is similar because, in most cases, management of email messages is on the local machine where the email is read. In contrast, with Web-based email, the central server, not the user's machine, manages the messages. These examples show how the style of data management can help classify the management as either distributed or centralized.

Centralized Management — Efficiently Using Resources, but Not Flexible

In many cases, centralized management for network services allows the most efficient use of resources, the greatest protection to its users from rogue applications[3], and the greatest ability to collect information about what users are doing. Unfortunately, centralized management often inhibits innovation because experimentation is difficult and more expensive to perform. A trade-off exists between the value of innovation and the value of a centralized controlling authority.

As discussed earlier, centralized management occurs when users cross organizational boundaries by accessing a server managed by someone unrelated to the user, or a centralized server manages the users data. Services with a centralized scope of management include 800 service, PBX service provided by Centrex, and Web-based email — the users belong to the general population while an unrelated party manages the service and data.

Centralized management of a network-based service offers several advantages from a business and technical standpoint, including efficient use of resources, efficient management, ease in tracking users, and transparency of function — users don't need to be engaged in managing the service at all. Generally, services with centralized management benefit from economics of scale due to the large potential user base. The PSTN is a good example of this; it serves its many users well at a reasonable cost. The success of the PSTN is a good argument for the benefits of centralized management.

With centralized management, knowing who the users are and what they are doing is much less complex. This has several benefits, including billing users, collecting accurate user data to leverage increased advertising revenue, and allowing better network resource planning and usage. The PSTN is a good example of this; the Central Office (CO) processes each call, and it records this information. The telephone companies have always been good at billing, partly because of their centralized network infrastructure.

Centralized management structure lends itself to efficient use of resources at every level. First, knowing who your users are and what they are doing is valuable information when trying to utilize scarce resources. Next, centralized management structure creates efficient management because people and equipment are likely at a central location. The consistency provided by centralized management promotes efficient training and help-desk service. Last, central management provides the most efficient use

---

[3] David Clark from MIT made this point in a Boston Public Radio show during a rash of Denial of Service attacks. He noted that the distributed nature of the Internet made it more vulnerable to this sort of attack than the more centralized structure of the PSTN, which has central information about each end-2-end connection.

of such physical resources as disk space. For example, in Figure 2.1 the centralized structure of the 800-number database[4] is far more efficient than the distributed nature of the DNS server. The SS7 network needs to send only a single query to find the real address of the 800 number[5]. This is unlike DNS, which may require messages to be sent up and down many levels in the namespace tree to resolve a name. The DNS may use caching to reduce this, but the caching adds to the complexity of the implementation and still requires traversing the tree when the cache is stale. From high-level resources, such as managing people and equipment, to lower-level resources, such as network usage, centralized management offers advantages in its efficiency.

By restricting what users can do, centralized management protects against the dreaded Denial of Service (DoS) attacks popularized lately in the press, partly because the central authority can track what users are doing. Knowing what users are doing also allows mechanisms for protection against rogue applications such as Napster (which is itself a centralized application). The ability to monitor traffic at a centralized location may allow easier detection and back tracing of any attacks. With centralized management, it is easier to discover attacks and abuse by applications.

Because networks with centralized management, such as the PSTN, know what their users are doing, it is easy to provide Quality of Service (QoS). Networks such as the PSTN will not allow a user to establish a new connection with guaranteed bandwidth unless the resources exist to provide the connection. In the PSTN, QoS is a byproduct of the network infrastructure. The centralized management of ATM networks with its predefined VPI/VCI is similar to the PSTN and allows absolute QoS. Centralized management supplies the information needed to provide QoS when resources are scarce.

In addition to the business advantages of centralized management, there are also technical benefits, including the following:

- Less complex feature development because of centralized information

- Less processing overhead because of centralized processing of data

- Less complex security because fewer locations and systems need to be secured

---

[4] There are several (7 or so) 800-number servers that are replicated for reliability and load sharing, but the lookup goes to only one of them.

[5] The 800 server may give different numbers located in different areas depending on the time of day.

For example, consider the example of the 800-number lookup as compared to the DNS service in the Internet. The single lookup request over the SS7 network is far less complex than a lookup in the DNS that must travel up a hierarchical tree and back down[6]. I will discuss other technical advantages to services such as email in Chapter 8 and basic/advanced voice services in Chapter 9. Because centralized management has both business and technical advantages, I group these together, calling them Business and Technical Advantages (BTA).

The benefits of centralized management are not free. In many instances, services with centralized management are not flexible to change, are difficult to experiment with, and do not give users much control over the service. Centralized network infrastructure serves many users; thus, any change to it has large potential impact. Chapter 4 discusses in detail the reasons why experimentation is so difficult with centrally managed services. A centralized structure allows for detection and tracking of hackers, which provides a real security advantage; however, the centralized management also gives hackers a central target on which to concentrate attacks — a security disadvantage. Another problem related to security is anonymity. While service providers view knowing what users are doing as a benefit, users may want more privacy. A good example of this is Web browsing; many users do not want a third party tracking and recording the sites they visit. While the advantages of centralized management are many, so are the disadvantages, making the choice between management styles a difficult one.

Distributed Management — Inefficient Use of Resources, but Flexible

In contrast to centralized management, a distributed structure provides a flexible environment that promotes experimentation, but it often lacks efficiency. In a service with distributed management, the servers are typically within the organization. The management of the servers is at the organization, department, or even desktop level. DNS is one example of a distributed system. At Harvard, this service is managed at both the organization level, with a top-level Harvard DNS server, and at the departmental level because some departments, such as Electrical Engineering and Computer Science (EECS), manage their own DNS. This serves Harvard's needs well, providing a service managed by the organization as a whole, but with flexibility, allowing a department to provide a local DNS service and thus

---

[6] If DNS caching is used, it becomes even more complex.

control its own namespace. This type of flexibility is what allows innovation with distributed services.

As a rule, services with distributed management manage the data and software in a distributed fashion, as in the DNS example. ISP Internet email systems using POP are another example of distributed data management because the local host running POP must retrieve the messages from the server and then manage them locally. Software may also have distributed management; desktop applications are a good example of this because each user controls his or her own software and the updates to it. This distributed management of data and software gives the user more flexibility, but at the cost of efficiency.

There are many advantages to distributed management, including more control by the user and easy experimentation. Sometimes users want control of their destiny; they want to decide when to upgrade a particular service, rather than depend on a service provider to upgrade hardware or software on a uniform schedule. This is the case when considering a PBX or Centrex; the PBX choice lets users plan upgrade scheduling, but with Centrex, the phone company decides when to upgrade the CO switch capabilities. User control and more innovation are two significant advantages to distributed management.

As expected, distributed management of services has disadvantages, including inefficient use of resources and difficulty of tracking users. Distributed systems require skilled help in many locations. A small company may not be able to justify hiring an expert with a particular system. It is more difficult to track users under a distributed management system because there is no central point of control. This presents a logistical obstacle to accurate billing. What if everybody received email by running email servers on their desktops? Counting the total number of email messages would be very difficult technically. A single mail server for everybody simplifies usage accounting. The distributed nature of the Internet makes it vulnerable to Denial of Service (DoS) attacks. It is these advantages and disadvantages of distributed management that make the choice of management structure unclear.

Distributed networks such as the Internet are finding it difficult to guarantee hard levels of QoS. Having no control over users — to the point of not even knowing who they are — makes it hard to allocate resources when they are scarce. QoS is not a natural byproduct of the architecture in distributed networks. There is hope, though; the Internet Engineering Task Force (IETF) has standardized technology for QoS within the Internet, and the schemes it has blessed offer a choice of management structure. The Internet has flourished within the business community despite the lack of QoS — a fact that many telephone companies still have a hard time believing.

One particular type of distributed management structure that is receiving much attention today is the end-2-end argument (see the next chapter). Services with end-2-end architecture by definition have a distributed structure because they push complexity to the end points of the network. The network is kept simple, with complexity built into the end, or edges, of the network. Applications that are end-2-end are unknown to the network infrastructure. This means that permission to add new end-2-end services is not necessary because nothing within the network knows about the new service. The end-2-end argument is one of increased innovation, and the proof of its validity is the success of the Internet in this respect.

Shifts in Management Structure

Whether a management structure works best, given a shift to the better architecture is never an all-or-nothing proposition. In most cases, it should be expected that similar services with different styles of management structure will coexist. The success of one management structure compared to the other is measured by the ratio of the market captured by each structure. I expect this ratio to shift as the market uncertainty changes. This book's theory expects that as market uncertainty decreases, centralized versions gain in market percentage, but as market uncertainty grows, distributed management structure becomes more popular.

The case study of email in Chapter 8 shows the coexistence of services providing a similar service using different management structures. In email, there is a broad continuum of use from distributed to centralized. Users who demand complete control of their email have their personal email server running on their desktop — a very distributed style of email management. Other users choose to use their ISP as an email server provider — a more centralized structure, but still having a distributed component. Most email growth today is with more centralized Web-based email services such as Hotmail or Yahoo!. The email case study shows that one management style does not meet all users' needs.

Similarly, the case study of voice services in Chapter 9 illustrates a continuum of use from a distributed to centralized management style. There will always be a market for state-of-the-art on-site PBXs managed by experts within the organization. Some users demand the control of owning and managing the equipment. Other users may require specialized features not yet implemented in Centrex, such as linkage to external computer databases for advanced call routing. An increasing number of users are opting for a centralized style of management offered by telephone service providers with Centrex. The centralized management inherent with Centrex frees many organizations to focus on their critical success factors, not on providing mundane voice services within the organization. How companies provide basic and advanced voice services within their organization illustrates how different management structures coexist for popular services.

Past approaches have not been able to explain why shifts in management structure occur when they have, and they are unlikely to explain when shifts will occur in the future. The problem with previous approaches has been that they did not consider when being able to innovate is more valuable than other business and technical advantages. This methodology helps explain why and when shifts in management structure transpire. It illustrates the importance of market uncertainty in shifts of management structure. By linking the value of experimentation to market uncertainty, this approach illustrates how to consolidate previous methodologies by factoring how market uncertainty affects the value of experimentation.

When does the value of innovation exceed the benefits of efficient management and resources? Previous approaches are partly correct: It just depends on the market uncertainty. When you are having difficulty meeting user needs allowing them to experiment has great value, but if the market is predictable, experimentation is not worth their effort. As Chapter 6 shows, in low market uncertainty the difference between the best and worst of many experiments is small.

To illustrate the preceding concepts, think about the Internet compared to the PSTN in the context of innovative new services. The Internet is dumb, and the PSTN is smart (see Chapter 3 for detailed definitions of dumb and smart networks), yet the innovation of new services in the Internet dwarfs those in the PSTN. The Web and the innovation it allows have far exceeded the services in the PSTN, partly because of who can innovate. In the Internet, anybody can, but in the PSTN, nobody can but the telephone company. It is a question of centralized control versus freedom for end users to create their own services.

Figure 2.2(a) shows a dumb network, such as the Internet. The nodes through which information flows do not know anything about the end users of the data. When Mark and Scott converse, the nodes the data flows through do not know what Scott and Mark are doing. In technical terms, the nodes keep no state about Scott and Mark. Figure 2.2(b) illustrates a smart network, such as the PSTN. In this intelligent network each node that is part of a connection knows about the connection, and each node keeps state information about each connection it has. The smart network knows what its users are doing, allowing it to protect its users, efficiently manage its resources, and unfortunately also impede users from innovating better network-based services.

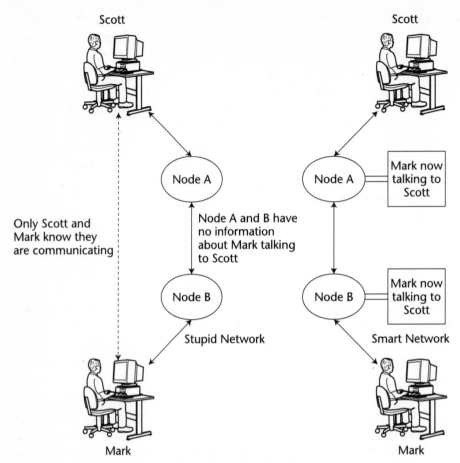

Figure 2.2   (a) Smart compared to (b) stupid network.

There is no agreement about what architecture works best to manage network-based services. Traditional telephone companies believe in the centralized model of the PSTN. Many "bell" heads still believe that the Internet does not work because it can offer only best-effort service. Others believe distributed architecture like the Internet is not able to offer robust network-based services because they are more vulnerable to security attacks such as the Denial of Service (DoS) attacks that have disabled major

Web sites in the past. There is no general right answer as to what works best; it depends on each particular case and on the market uncertainty.

This question of management structure of network-based services transcends the particular network in many cases. For example, as discussed, within the PSTN the PBX offers a distributed structure as compared to the centralized managed structure of Centrex. In the Internet several architectures provide Voice-over IP. SIP and H.323 allow a distributed model, but megaco (H.248) does not. This shows the complexity of the decision about network infrastructure and network-based services. Distributed networks, such as the Internet, and centralized networks, such as the PSTN, both allow distributed and centralized architecture for some types of network-based services, making the decision about which is best very complex.

Conclusion

This chapter discussed network-based services and management structure. The value of experimentation with network-based services is highlighted. Management structure was explored in the context of how it affects the ability to experiment with innovative network-based services. Some network infrastructure, such as that within the Internet, tend to allow users to experiment, while other architectures tend to discourage users from innovation. Distributed management promotes innovation by allowing easy experimentation, but it may not be efficient and may have other technical disadvantages. On the other hand, centralized management is efficient and has other business and technical advantages, but it does not encourage innovation and may even disallow users from experimentation. Choosing the management structure that works best is a difficult decision.

In the next chapter, the end-2-end distributed type of management is explored in more detail. The end-2-end model is an icon of historic Internet design. The next chapter discusses this argument in its historical context and in the context of the new Internet. It presents a way to think about the distributed management structure of any process compared to centralized management.

CHAPTER 3

Internet End-2-End Argument

The end-2-end argument is very important to network and system designers, managers, and investors because it is an example of how a technical
design principle affects the very nature of what users can accomplish. This
argument promulgates the idea of networks with simple internal structures (the stupid network discussed in Chapter 2) that are unaware of what
end users are doing. It proposes that the complexity of applications should
be at the end or edges of the network, not within the network. The Internet
has historically promoted end-2-end structure and has flourished because
of the innovation this structure has made possible. Managers who understand the end-2-end principle have a competitive advantage because they
are better able to structure the management of network-based services that
will more likely meet uncertain user needs. Similarly, investors grasping
the end-2-end argument have an advantage because they have another tool
to value investment strategy. The end-2-end argument provides a way of
thinking about how to structure a system.

   This chapter defines the end-2-end argument and offers reasons why
managers and investors should learn to think about the network (or other)
infrastructure with the end-2-end point of view. The chapter will help you
apply end-2-end principles by understanding how market uncertainty
increases the value of end-2-end structure. This chapter explores the history of the end-2-end principle from its start as a technical design principle.

Next, the end-2-end argument is discussed in the context of the way the Internet is changing and why/how the end-2-end argument is mutating to better fit the current Internet environment. This chapter points out how some current network applications such as Network Address Translation have potentially devastating consequences to innovation in today's Internet.

Why is the End-2-End Argument Important?

The end-2-end principle is a large part of the design philosophy behind the Internet [1]. According to the end-2-end principle, networks should provide only the simplest of services. The end systems should have responsibility for all applications and any state information required for these applications. By providing the basic building blocks instead of complex network services, the network infrastructure will not constrain future applications. These arguments started in the 70s in the context of how to encode and send secret encrypted messages with secure and efficient processing; then other applications such as transaction processing used them to justify their design. Later, the end-2-end principle became a more general rule guiding the design of the Internet's infrastructure. The changing nature of the Internet has caused a renewed interest in this old argument.

A manager who understands the end-2-end principle and how and when to apply it has the advantage of being able to tailor the environment of network services to the needs of a particular user base. For example, consider how to manage a network file system along with the applications running on a particular desktop. What should the user be able to do? Are the main applications common to all users? Should a user be able to have their own applications? The end-2-end argument would suggest that a network file system should be as simple as possible. It should only provide files across the network and allow users to do as they please in the context of their files and applications. Allowing users to do as they please, though, has disadvantages — they may install software with viruses. It is also difficult and expensive to manage applications at the desktop when many users are running different versions of that application. The argument linking the value of the end-2-end principle with market uncertainty illustrates the value of allowing users to do as they please when you can't seem to please them.

Investors in technology, network service infrastructure, or service providers that understand when end-2-end architecture creates value have an important way to approach a business deal. This end-2-end idea argues that large centralized service providers, such as traditional telephone companies, cannot meet uncertain markets with the services they think up themselves. It suggests that having many small service providers and users that can experiment will produce more innovative business solutions. Applying the end-2-end principle correctly does not imply the demise of big centralized service providers, but it does change their business models. It suggests that large centralized service providers need to implement that which is already known to meet users' needs. The end-2-end argument illustrates to investors that a less efficient structure might still be worth more because users will be better satisfied.

One reason the end-2-end argument is relevant today is that new services, such as Voice-over IP, have a choice of management structures allowing voice-based network services to have either an end-2-end architecture or a more centralized model. Session Initiation Protocol (SIP) is one way to provide Voice-over IP. SIP can provide true end-2-end service, with nothing except the end points knowing that a conversation is taking place. Implementations of SIP work, but it is not the favorite model of traditional telephone companies. The protocol known as H.248 (also known as megaco) is in codevelopment by the IETF and International Telecommunications Union – Telecom Standardization (ITU-T). Based on the current architecture of the telephone network, it is a more traditional way to provide Voice-over IP. It relies on a centralized server that coordinates the voice connections. This centralized server has knowledge of all the calls it processes — a nice feature for billing. It is not surprising that traditional telephone companies support the H.248 (megaco) model; they are sure it is superior to SIP. The disagreement about how to provide Voice-over IP illustrates the importance of the end-2-end argument in today's Internet because it helps to frame the problem.

Definition of End-2-End

One idea that has helped the success of the early Internet is that its infrastructure permitted early end-2-end applications, such as Telnet and FTP. Services with end-2-end architecture [1] by definition have a distributed structure because they push complexity to the end points of the network. The idea is to keep the network simple and build any needed complexity into the end, or edges, of the network. Applications that are end-2-end are unknown to the network infrastructure because there are no application-specific functions in the network. This means that changes to the network or permission to add new end-2-end services is not necessary because nothing within the network knows about a new service. The end-2-end argument is one of increased innovation, and the proof of its validity is the success of the Internet with regard to innovation.

One example illustrating what end-2-end services look like compares hop-by-hop to end-2-end encryption of data [2][3], as depicted in Figure 3.1. In the hop-by-hop case, the network node in between Bob and Alice must understand what Bob and Alice are doing. This transit node (Tom) must have knowledge about the encryption application that Bob and Alice are using. This node in the middle might be independent of both Bob and Alice, but Bob and Alice must keep it appraised of the application they are running. This makes the service not end-2-end. In this hop-by-hop case, Bob encrypts his message in a key common to him and the node the message must transit in its journey to Alice. Tom decodes the message in the key shared with Bob, and reencrypts the message in a key shared with Alice (which is likely different from the key the transit node shares with Bob). Finally, Tom sends the message to Alice in the key it shares with her. This hop-by-hop method is not secure because Tom knows how to decode the message. The end-2-end case is less complex. Bob encrypts his message in a key shared only with Alice, and the transit node cannot look at the message. Tom must only receive the message from Bob and forward it to Alice. This end-2-end methodology is more secure because nobody in the middle of the network can view this message from Bob to Alice. In the end-2-end case, only Bob and Alice need to know what they are doing. The node in the middle knows only that it must forward any data from Bob to Alice, and from Alice to Bob. This application independence of Alice and Bob to the network infrastructure is what makes the service end-2-end.

The advantages of end-2-end encryption include the following:

- Increased security because no transit nodes are able to decode the encrypted message. Trusting a third party should be problematic to Bob and Alice. Intuitively, it makes no sense for Bob and Alice to trust someone else.

- Less processing overhead because the transit node does not need to decode and then reencrypt the message. Both encryption and decryption require extensive computer resources.

- Ability to change and innovate because Bob and Alice can do what they wish without informing any transit nodes. This includes changing encryption/decryption keys and invoking new security methods at their own discretion without informing the manager at the transit node. This enhances the ease with which Bob and Alice can experiment and innovate because they need only each other.

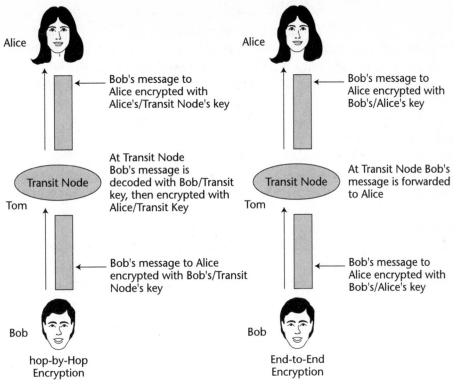

Figure 3.1 End-2-end encryption service.

In the preceding case it is obvious that the end-2-end method is better in all aspects: It is more secure, efficient, and flexible. Don't be fooled by this example; there are many cases of network-based services where the choice of whether end-2-end architecture makes sense is not so simple. In the past as well as at present, distributed management of network-based services, such as email and voice, makes sense in some cases, even though the cost of providing the end-2-end architecture exceeds that of a centralized management structure.

The end-2-end argument states that services offered by the network infrastructure should be as simple as possible. If you try to anticipate the services applications will need, you may be wrong, and you will most likely inhibit new applications by constraining them with services that do not match their needs. Networks that provide only simple, basic services allow applications more flexibility in what they can do. The IP protocol in the Internet is a good example of this philosophy; it is simple, offering just the most basic type of network service — the unreliable datagram service.

This simple core protocol has allowed immense innovation at the transport and application layers. Different application modules can utilize different transport protocols that match their needs, but all of them are built over IP, which has become the glue holding the Internet together. The success of the Internet is partially due to the simplicity of IP. Again, this validates the end-2-end argument.

By pushing applications to the user level with end-2-end applications, more experimentation is likely. There are several reasons for this. First, application-layer development is faster and less expensive than kernel work because kernel code tends to be complex and debugging is often difficult. Next, the pool of talent with the skills to do application-layer coding is greater. Finally, those programmers allowed to develop new services are much broader at the application level because they include users, and as Hippel [4] shows, users sometimes are best suited to solve their own problems.

Because end-2-end applications do not require network infrastructure change or permission to experiment, users can and do innovate new services. Consider the creation of the Web. Tim Berners-Lee [5] was not a network researcher searching for innovative ways to utilize the Internet, he was an administrator trying to better serve his users. He developed the World Wide Web to allow the scientists in his organization to share information across diverse computers and networks. It just so happened that his solution, the Web, met many other user needs far better than anything else at the time. This illustrates one powerful attribute of the end-2-end argument — you never know who will think of the next great idea, and with end-2-end services, it can be anybody.

History

As described, the end-2-end argument [1] has been around since the 70s in the design of message encryption systems [2][3]. Later, other applications such as transaction processing used it to justify their design. It became one of the underlying principles of the Internet's infrastructure, and it is a main reason that innovation has flourished within the Internet environment, but not in the traditional telephone network. Understanding how the end-2-end argument has helped the Internet become successful will help managers understand how to best structure the IT infrastructure and then determine the optimal management architecture of services built on this infrastructure.

When David Reed was a graduate student at MIT in the 1970s, he was thinking about transaction processing and how to architect a robust and reliable protocol to allow updating of data in a distributed environment. He was trying to decide what services a network should provide for application data update protocols. His answer is one early and important step in the history of the end-2-end argument. In Chapter 2 of his thesis [6] about a two-phase-commit data update protocol, he elegantly argues that networks can lose data, can deliver data in a different order than it is sent, and can even duplicate data, and that a robust update protocol must work with all of these errors. He argues that, even if the network provides perfect data transfer, the two-phase-commit application must still perform these tasks by pointing out that faulty application software or bad hardware can cause all of the preceding symptoms. He concludes that, because the ends must perform these tasks, the network need not worry excessively about them. This is the essence of how the Internet's network protocol IP works: It is unreliable, and data can be out of order or duplicated. Reed's thesis explains why simple networks and complex applications are a good combination.

Reed's seminal argument is extended to explain why networks should provide only simple basic network infrastructures, in a paper [1] by Saltzer, Clark, and Reed. They argue that it is impossible to anticipate the advanced services new applications will need, so providing just basic services is best. They point out that trying to meet the needs of unknown applications will only constrain these applications later. The paper argues that simplicity of basic network services creates flexibility in application development. Complex networks may allow easy development of certain types of applications (that is, those envisioned by the network designers) but hinder innovation of new applications that were beyond the grasp of these same network designers. The principle of simple networks was at odds with the belief of the telephone companies at the time, but fortunately it became a major influence in the design of the Internet.

David Isenberg is an interesting fellow. While working for a telephone company (AT&T) that was spending billions of dollars designing and implementing the next generation of intelligent networks, he was writing his classic article about the dawn of the stupid network [7]. He discusses problems of intelligent networks (such as the telephone network) and explains the advantages of networks with a stupid infrastructure, such as the Internet. He explains that some of the technical advantages of simple networks are these:

- They are inexpensive and easy-to-install because they are simple.
- There is abundant network infrastructure because it is inexpensive to build and maintain.
- They under specify the network data, which means they do not know or care about this data.
- They provide a universal way of dealing with underlying network details (such as IP).

Isenberg discusses how user control boosts innovation. One of the major contributions of this work is that it is the first to mention the value of stupid networks in the context of the user's ability to perform experimentation at will and share the results with friends and colleagues. His work was well received by many, but not by his management, who believed that the stupid Internet could never satisfy the demanding needs of the business community. It turned out that management at AT&T was wrong, and David Isenberg was right. He is now a successful independent consultant, writer, and industry pundit.

Another important contribution to the end-2-end argument was made by the author in his thesis [8] and in work with Scott Bradner, Marco Iansiti, and H. T. Kung at Harvard University. It contributes several ideas to the end-2-end principle, by linking market uncertainty to the value of end-2-end architecture. It expands on Isenberg's ideas about the value of user innovation by explaining that allowing users to innovate creates value because there are many more users than service providers. Furthermore, because it often costs less for users to experiment, they may perform more experiments, thus increasing the expected value of the best of these experiments. The major contribution of this work is that it links the level of market uncertainty to the value of user experimentation and innovation. User innovation is of little value if market uncertainty is low because the service provider will create services that meet user needs as well as anybody. It is likely that every proposed service will meet the needs of a certain market. Furthermore, the big centralized managed service providers will use resources more efficiently. It is when market uncertainty is high that user innovation has the greatest value because the larger number of experiments increases this value. This illustrates the link between market uncertainty and the value of end-2-end architecture.

The creators of the Internet believed in the end-2-end argument, and the basic Internet network- and transport-layer protocols IP, TCP, and UDP are examples of its application. The network-layer protocol IP guarantees little; data can be lost, reordered, and repeated. IP, however, is very flexible and allows different types of transport-layer protocols, such as UDP, TCP, and now SCTP. These different transport protocols built on the simple IP layer give applications the flexibility they need by allowing them to choose a transport protocol suitable to their needs. The network should not decide the type of transport for the applications. Some applications, such as DNS, found unreliable data service worked well; other applications, such as HTTP or FTP, need reliable data service. Different applications will demand different services, and the network should not constrain these choices. The end-2-end argument helped the designers of the Internet promote the development of applications by users because of the flexibility it gave developers.

End-2-End in Today's Internet

Today's Internet is different from the one from early days because of the scope of its use and those who use it. The end-2-end argument that helped the Internet become great must adapt to the current reality of the Internet. Clark's and Blumenthal's [9] recent paper on rethinking the design of the Internet shows how its basic communications requirements have changed because users do not trust others and the corporate world has connected to the Internet. They discuss the impact of new network devices such as Network Address Translators (NATs) and firewalls that affect the end-2-end nature of Internet applications. They show how such devices reduce innovation in the current Internet by limiting what users can do. Network services are discussed in the context of the new Internet and how they fit into the end-2-end argument. A service that authenticates a user before allowing him or her to perform a task is not end-2-end in the pure sense, yet it is a combination of end-2-end transactions. Even more important, this type of verification from a "trusted authority" makes intuitive sense in today's Internet. This work points out that some alterations from pure end-2-end services still maintain the general idea, but other alterations, such as NATs, are not justified and break the end-2-end paradigm in unacceptable ways.

Along the same argument as Clark's and Blumenthal's is Carpenter's RFC2775 [10] (also look at Carpenter's RFC1958 [11]), which discusses the effect of network fog. Network fog are devices in the network that limit a clear end-2-end view of the network. Given the changing nature of the Internet, some things must change, such as centralized security services needed to allow e-commerce. Other ideas, such as non-unique global addressing made possible by NATs and popularized because ISPs limit the number of IP addresses per customer, severely affect the ability of users to continue innovation on the Internet. Carpenter is worried about the current direction of parts of the Internet infrastructure, as he should be.

Have you ever wondered how to hook up several computers at home to the Internet when your cable/DSL provider gives you only a single IP address? Unfortunately, the solution is to use a NAT at home, but this breaks the end-2-end model. NATs change the network address within the IP header of a data packet. This might seem harmless, but it is not. Using a NAT stops some services from working. For example, most Voice-over-IP protocols including SIP and H.323 break with NAT technology. MSN voice chat does not work with the NAT solution either because for these voice applications to work, the IP address of each end point must be known by each end, something not possible with a NAT. NATs also break many other Internet protocols, such as end-2-end IPsec security, and even basic file transfers using FTP. While not prohibiting innovation, NATs do make it more difficult.

The end-2-end argument is moving into the mainstream of telecommunications policy because lawyers such as Larry Lessig [12][13] have realized the importance of end-2-end structure and its relationship to telecommunications policy. There are some important policy questions that the end-2-end argument can help resolve. Of particular interest to Lessig is policy related to owners of cable networks and independent service providers that want to provide network services using the basic infrastructure of the cable network. Should cable network owners be able to control what services (and hence what content) a user can choose from? The end-2-end principle implies that this is not a good idea. Users must be able to choose the services and the providers of these services without interference from cable network owners. Any other policy stifles innovation. Today's telecom policy about open cable network access will sculpture tomorrow's landscape; we need to do it right. As the Internet shows, "right" means allowing end users to innovate today with broadband Internet services and letting users, not cable network owners, choose the services and content they want.

Conclusion

The end-2-end argument has provided, and will continue to provide, insight into what network services the network infrastructure should offer. The end-2-end theory is becoming more mature, allowing managers to understand when to apply the end-2-end argument in the context of how well they understand the needs of the users they are serving. End-2-end architecture has the most value when users have uncertain needs because it allows more experimentation. Knowing when there is value in letting users create their own services and when the cost of user innovation is too great allows management decisions that create the most value. The end-2-end principle is another tool for the network manager's toolbox.

Understanding the end-2-end argument is one methodology that investors can apply to value investments in network-based services. It may be that the benefits of centralized management will far outweigh the value of the flexibility allowed by end-2-end architecture. Alternatively, it may be that uncertainty is so great, efficiency is of little value. Knowing when the value of distributed end-2-end architecture is worth more than cost-efficient centralized structure helps an investor decide when to invest in services based on their architecture and the level of market uncertainty.

The next chapter examines the difference between distributed (end-2-end) and centralized management. It provides more detail about the advantages of distributed management, such as easy innovation, and the disadvantages, such as inefficiencies in the use of resources. It further explains why centralized management is more efficient, and it offers other business advantages, such as protecting users from each other. It explains why centralized management is inflexible in the context of allowing users to innovate. It proposes a multilevel hierarchical structure to frame the process of deciding about how to provide network-based services within an organization. Examples are given from email, voice, and information services, illustrating how each of these services exists with many different management architectures, each structure meeting the needs of a particular user group. Whereas these services are different, a common thread runs through the examples because of how market uncertainty affects the decisions about management structure. It illustrates the complexity of the decision between centralized and distributed management structures.

Management Structure of Network-Based Services

Understanding the link between management architecture and market uncertainty is critical to successful decisions about how to build a network infrastructure and services using this infrastructure. Network-based services coexist with a continuum of management structure from centralized to distributed, as discussed in Chapter 2. It is expected that experimentation will be easiest toward the distributed end of the spectrum and more difficult as the management structure becomes more centralized. Examples from email, voice services, and informational portal sites show that successful implementations of all these services coexist with different management structures. The different management architectures appeal to users with different business needs. Companies with critical success factors dependent on cutting-edge telecommunications services need more flexibility than centralized management allows, but organizations that have critical success factors less related to technology find centralized managed services a good fit with their needs because of its efficiency. The link between management structure and market uncertainty helps investors better structure their investment strategy and managers build network infrastructure and services creating the most value.

Centralized versus Distributed Management Structure

Network-based services have many levels of management architecture within a company. First, a company must choose whether to manage its own services (a distributed architecture) or outsource the service (a centralized structure) to an independent service provider. If the service is outsourced, then there are other decisions to make related to the degree of centralization of the outsourcing. If the decision at the top is to self-manage a service, then the architecture of network infrastructure used to provide this service must be decided. At each level in this hierarchical structure market uncertainty should guide the decisions about how centralized the management structure should be.

Each organization has a different level of market uncertainty for each of the network-based services it uses. This market uncertainty depends on the technology and the needs of the company, as determined by methods from information technology strategy, such as critical success factors, Porter's five-force analysis[1], value chain analysis, and cost-benefit analysis [1]. These methods help decide what to build and what business needs the system must meet. Each organization will have unique needs, and thus the correct mix of management structure for the host of network-based services each company requires will be different. Low market uncertainty indicates the commodity-like services implicit with central management architecture will work; high market uncertainty implies that the choice of the many cutting-edge applications made possible by distributed management structure creates more value. This chapter is not a recipe to follow, but rather a way to think about how to create the appropriate unique blend, by considering how market uncertainty affects the ability to meet user needs within the context of the management structure.

This chapter examines the link between management structure and the ability to experiment in order to meet uncertain user needs. First, a generalized argument is presented about the difficulty of experimentation with centralized management, and it also explains how a distributed management structure promotes experimentation. Different styles of management architecture allow the organization using the structure to have varying degrees of innovation. Email, voice, and Web-based information services are examined in the context of how their management structure changes the level of experimentation expected with each style of management architecture. Each of these services is discussed in the context of its coexistence

---

[1] Porter [5] identified five forces that affect the strategic direction of a company. These forces include strength of the suppliers, strengths of the customers, threats of substitute products, barriers to entry, and rivalry.

with similar services using different management architectures, how each architecture meets the needs of a particular user group based on the market uncertainty of that group, how each style of management affects how much experimentation takes place, and the value of this experimentation.

As discussed in Chapter 2, centralized management has many advantages, such as the following:

- Economies of scale that allow more efficient use of resources
- Knowing who your users are, and what they are doing
- Protection of users from each other
- Easier billing
- Allowing the organization or department within an organization to focus on its core competencies

Centralized management also has disadvantages, such as these:

- Inflexibility to change and difficulty with experimentation
- Providing a single point of attack for hackers to focus their efforts on
- Difficulty for users in remaining anonymous
- Allowing a single point of failure
- Changes that impact a wide range of users

The advantages to distributed management include the following:

- Users have control over upgrades and maintenance schedules.
- Users can remain anonymous if they choose and have the technical skill.
- Experimentation is inexpensive and easy — anybody can do it.

The disadvantages of distributed management include the following:

- Less efficient use of limited resources
- Difficulty in billing users because you might not know what services they are using
- More skilled people needed to manage the service
- Difficulty in protecting users from each other

The next section focuses on the flexibility of a management structure within the context of innovation and experimentation. It illustrates why experimentation is hard to do with a centralized management structure and easy to do with a distributed architecture.

Why Centralized Management Increases the Difficulty of Experimentation

Centralized management structure imposes a centralized controlling authority in charge of management. This manager weighs the needs of all users across organizational boundaries and attempts to ensure that the greatest number of user needs are met to the greatest possible degree. This approach, however, makes the service less tolerant to the changing needs of any particular user. These managers view changes with fear because any modification could interrupt service to many users. A bad experiment is a potential disaster because it may affect a large number of users across many organizational domains. The very nature of a controlling authority works against the type of random experimentation that is valuable when market uncertainty is high.

One essential question about experimentation is this: Who is performing the experiments? Are the experimenters highly trained (and possibly myopic) professionals with advanced degrees and years of experience, or are they a combination of highly trained experts, creative people without formal training, and the clueless? Large centrally managed service providers, such as the local phone companies, allow only highly trained, experienced workers to create software for new services. These individuals tend to be conservative in their approaches, their time limited by schedules and the plans of upper management. The only ones allowed to experiment with centralized services have too many time constraints, and they have no incentive to innovate.

The controlling central authority of a network must give permission to change the infrastructure within the network. Consider popular *## services in the PSTN such as *69, which retrieves the number of the last incoming call. Adding a new service or changing an existing service requires a network change, but no change to the end device. As discussed previously, changes to the network infrastructure are difficult and expensive. In addition, changes to the infrastructure require the permission of a central manager who is not inclined toward wild experimentation. The mechanisms to allow changes to the infrastructure will likely involve this manager, who is more inclined toward safety than experimentation that has any likelihood of failure. The difficulty of receiving authority to change the infrastructure within a network is one reason that centralized management inhibits innovation by limiting experimentation.

Sometimes the complexity of systems with centralized management is high, causing experimentation to be costly. The software controlling features to residential users within the phone network has this property [2].

The potential conflict between features can be very complex. For example, consider the interaction between caller ID and a busy phone line. At first, a user could not receive the caller ID for an incoming call if he or she was on the phone. Bellcore needed to invent a highly complex protocol called Analog Display Services Interface (ADSI) [2] to provide this obvious service. This complex, combinatorial type of conflict between features grows rapidly with increased features and accounts for the high complexity and expensive experimentation inherent in large systems with centralized management structure.

Experimentation with new services on large centralized servers is risky. An experiment gone bad might impact a lot of users, maybe all of them. Managers of large centralized systems are wary of experiments and prefer stability. Suppose a new service is being phased in with an undiscovered bug that causes excess consumption of system resources, which results in the exhaustion of a globally critical resource. With a large centralized server, all users will experience Denial of Service in this situation, but with a more distributed architecture, a smaller group of users will be affected. There is a correlation between the number of users on a system and the cost of a failed experiment that affects all users of the system.

While efficient in its use of resources and offering other business and technical advantages, centralized management does not promote experimentation. The central control allowing the business and technical advantages is what inhibits innovation. Knowing what everybody is doing means that everybody will do less. Depending on market uncertainty, centralized architecture may or may not be worthwhile. The higher the market uncertainty, the less likely it is that a centralized management structure will work well because the difficulty with experimentation will make it hard to meet users' needs.

Why Distributed Management Promotes Innovation

Services with distributed management have no central authority controlling what the users do. Typically, users can do whatever they want, allowing them to experiment with new services. These end users can be conservative and highly trained or wild and crazy. One good example of the power of allowing users to innovate is the Web. Created by a user, it vastly surpassed anything that the so-called experts of the day could imagine. Even more unlikely, it came from Europe in the beginning of the 1990s, which is surprising because of the strong support for OSI at the time within the European community. The Web exists because of the creativity of one system manager trying to better meet the needs of his users. Users find that not having a central manager control their destiny is a liberating experience, one that creates opportunities to innovate services that better meet their needs.

One advantage of end-2-end services with distributed management is that the network does not know what applications users are developing or what services users are accessing. This makes it impossible to restrict experimentation because there is no way to determine its existence. This means network infrastructure does not need to adapt to new applications; it remains constant, and the applications use the simple yet flexible services it provides. This property of network infrastructure being application-unaware gives innovation a powerful boost because it simplifies the process of experimentation and allows users to participate in innovating new services.

With distributed systems, experimentation can be limited to only a few end systems, which is an advantage if the experiment fails. With fewer users feeling the impact, the experimenter has fewer worries and can focus on the experiment, not on protecting many users. Consider the following example: adding a feature to a mail server (Sendmail) to filter out spam email. The architecture of Internet email allows this addition without affecting any users except the one adding the feature. At first, I can run simple experiments and worry about only just local feature conflict. Only later, if the experiment is a success, do I need to make my new feature work in all cases. When making changes like this to a centralized email server such as Hotmail, it is hard to limit the users of the new change; thus, it must work with all features from the very start. Distributed management structure limits the scope of the impact of experiments and is a good way to promote innovation.

Distributed management allows those not managing the network to experiment with new services. On the distributed Internet anybody can try a new end-2-end experiment. If it works well, as the Web did, it might change the world. Because anybody can experiment, and because it is not expensive, the entry barriers to innovation are low; all you need is a good idea and some talent. That is how the Web was created. This type of wild experimentation, where users are allowed to innovate, has the most value when market uncertainty is high. It is the crazy ideas that have the most potential to be big winners.

While flexible and conducive to innovation, distributed management structure for a service can be far from efficient in terms of management and use of resources. Allowing users to do as they want makes it hard to best allocate scarce resources, know what users are doing, and protect these users. User freedom equates to management hardships. It is market uncertainty that helps the manager decide when the freedoms implied with distributed management exceed the business and technical advantages of centralized management. The higher the market uncertainty, the more likely the power to innovate with distributed architecture exceeds any business and technical advantages to centralized management.

Hierarchical Nature of Management Structure

The examples presented in the next section illustrate a common thread among services such as email, voice, and information — these services involve making choices about the architecture of the management structure at several different levels. The highest level is the decision to outsource the service or manage it in-house. The next level in this management structure hierarchy depends on the decision at the previous level. If self-management is the top-level choice, the subsequent decisions are different from those related to outsourcing. This is similar to the hierarchical nature of design that Clark [3] writes about when he describes high-level design choices in products such as automobiles, where the type of engine dictates the later design decisions such as fuel storage. This tree structure illustrates how a decision needs to be made at each level as to how distributed or centralized that level's management structure should be.

Trade-offs at each layer exist between the flexibility offered by the distributed structure and the business and technical advantages of more centralized architecture. It is the market uncertainty at each level in the hierarchy that helps determine what branch to take next in a path down the tree. This hierarchical framework is a way to organize the choices management must make in order to provide services by systematically examining the options at each branch of the decision process in the context of market uncertainty. If market uncertainty is low, then the centralized management structure makes sense; however, higher market uncertainty implies that the value of experimentation enabled by the distributed architecture exceeds the business and technical advantages of centralized management.

Figure 4.1 illustrates this abstract hierarchical nature of management architecture. It shows how the top-level architecture decision determines the next decisions that need to be made. The choices at each level of the tree are determined by the choices made above it, with each path down the tree containing a unique set of choices. Each decision, at each level of the tree, has unique market uncertainty associated with it. High market uncertainty suggests that the distributed structure of the right part of the tree (dotted lines) is best; low market uncertainty implies that the centralized management architecture of the left path (solid lines) is best. This tree structure is a logical way to map out the hard decisions about the management structure that need to be made based on the market uncertainty.

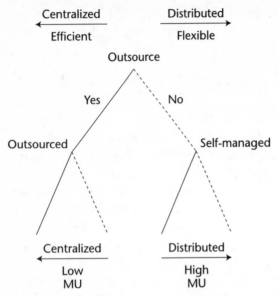

Figure 4.1 Hierarchical structure of management architecture.

Examples of Management Structure

Below, I discuss three different services that illustrate how management structure affects the ability to innovate. Managers must make choices when providing email, voice, and information portals that best meet the needs of users for these services. Should the service be outsourced or managed in-house? If managed in-house, should the internal architecture be centralized or distributed? In the next sections, these topics are discussed within the context of providing email, voice, and information services that meet user needs when these needs are uncertain. The analysis of these examples illustrates the real options way of thinking proposed in this book and explained with more detail in Chapter 5.

Email

Today, email is a requirement for most companies. It has become fundamental to business, family, and personal life. Email today works because everybody has agreed to the Internet set of standards, which implies interoperability and gives economics of externalities [4]. No matter what service or system the organization chooses to use, it will be able to communicate with everybody else. This standardization has created many choices for users because there is no best management structure to provide this service.

It depends on the organization and what factors are critical to its success. At one end of the spectrum, a company can outsource its email to a large centralized Web-based provider, such as Hotmail or Yahoo! — a very centralized solution. Many users (such as myself) have Hotmail accounts that are useful and free — a powerful incentive. If Web-based email does not meet an organization's needs, outsourcing to an Internet Service Provider (ISP) might. On the other hand, sometimes users demand complete control over their email systems, meaning they will self-manage their email service. If email is provided in-house, then the network infrastructure might be centralized with a single email server or distributed with many email servers. There are many ways to provide this necessary service, and each structure makes sense for some groups of users, depending on the amount of market uncertainty within the group. Each different way to provide email service has its pluses and minuses — the level of control each user has, the ability to experiment, the cost of experimentation, and the ease of management differ drastically from one extreme management structure to the other. With centralized management structure, features such as virus protection and filtering email for inappropriate language are easy to manage, but management becomes more difficult with distributed management structure. The case study in Chapter 8 illustrates that when market uncertainty decreases, the percentage of users satisfied by email systems with more centralized management architecture increases.

The first decision about how to provide email is choosing whether the organization wants to outsource this function. While virtually all organizations need email, for some it's like a commodity; they need only the most basic services, such as a stable email address. A centralized Web-based email provider, such as Hotmail or Yahoo!, or a large ISP, such as AOL or MediaOne, will meet their needs well. Outsourcing lets a company spend the least time managing email, allowing it to focus on its critical business functions. If you are certain about what you need, and if you are sure the email service provider can meet your current and future email needs because you expect ongoing low market uncertainty, then outsourcing could be a good idea.

Once you decide to outsource email, there are additional choices to make, as illustrated in Figure 4.2, which shows several ways that email might be outsourced. In part (a) both Companies (B) and (C) outsource their email service to a large centralized Web-based email service provider, such as Hotmail or Yahoo!. This is the most centralized way to provide email. With Web-based email, the small business owner needs access only to the Internet, because from anyplace in the world with Internet access and a browser, users can access email. Depending on how much email the user wants to store on the centralized server, a service like this might be free or have just a small fee. The local company has no responsibility to manage the mail server or even the database of email messages. The service provider manages these tasks. The email messages themselves are managed by the email service provider, which means that if the service is down, the user cannot even access old email messages. With this email system, users have little flexibility. If they are unhappy with the service or demand advanced features not currently implemented, it is unlikely that they will get any satisfaction. In this case, the power of the user is very low (see Porter's five forces [5]) because it is unlikely that any one user can get large centralized Web-based email service providers to respond to any request for different service. The centralized scheme depicted in (a) is the least trouble to manage for the organization, but it is also the least flexible, because the users have no control over how the service provider manages the service.

In contrast to Web-based email is the outsourcing methodology shown in Figure 4.2(b). With this email system, the organization has outsourced its email service to an ISP such as America Online (AOL), MediaOne, or smaller local ISPs, via the Post Office Protocol (POP) or Internet Mail Application Protocol (IMAP). When the email system is outsourced to an ISP, the user does not need to manage the email server, but most likely will manage his or her individual email messages. It depends on whether POP or IMAP is used (IMAP is not widely implemented with ISPs). With POP users have only one choice — to manage the messages themselves; with IMAP users decide if they or the service provider will manage the messages. Providing email with this method has a centralized structure, but less so than the Web-based email scenario. There are two main reasons for this. First, ISPs have more of a membership structure. This is true even for large ISPs such as AOL and MediaOne — if you are a member of one, then that is where you get your email. Members of both AOL and MediaOne can get email service from a large centralized Web-based email provider such as Hotmail. Second, ISPs mostly force users to manage their own email messages. The less centralized email scheme in (b) is easy to manage, provides little flexibility, but does allow the user more control over his or her email messages if the user chooses that option.

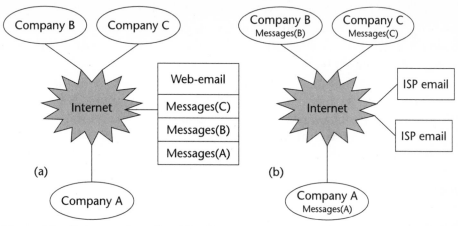

Figure 4.2   Outsourced email architecture.

As seen previously, one decision a manager faces is who should manage email messages — the service provider or the user? When the messages are managed by the service provider, the user has no responsibility for message backup, disk space, and database management. When the user manages his or her own messages, then the user has this responsibility. Both schemes work for different sets of users. If you are not sure what you want to do with old email messages (an indication of higher market uncertainty), the flexibility of self-management is likely worthwhile.

Different organizations have different needs when managing messages. A company might depend on fast retrieval of old deleted email messages from backup archives. Depending on an ISP or Web-based email provider to restore email messages just won't work; the company must have access to the information, and the only way to do this is by local email management. Some organizations, such as stockbrokers, are required by law to archive old email messages and maintain their own email backups. Maybe a company does not want any records of email, which makes backups by ISPs and Web-based email providers unacceptable. Remember the

Microsoft trial and the embarrassing emails that surfaced in court; this is a good argument for not archiving all email, a strategy that can be ensured only with a self-managed email system. When you need control and flexibility, running a mail server seems like a good idea.

If email is essential to the success of an organization, then outsourcing email may not be the best answer because such organizations demand control over their email services. This also depends on whether the particular organization has the technical ability to provide a more robust and flexible self-managed email system. Maybe the organization wants to decide when to upgrade new features or wants to be able to monitor the content of email messages or choose the way messages should be archived. Providing your own mail service by running a mail server within the organization gives management the flexibility to do what it wants, when it wants. For these organizations, the hassle of managing a mail server is worth the effort because they are unable to meet user needs any other way.

There are many ways to structure email systems within the organization, some of which offer more flexibility than others. The organization might run a centralized mail server for the entire organization. All email into and out of the company must go through this centralized email server because it is the only server in the organization. This approach is easy for the organization to manage and gives it the most control over how the system is used. On the other extreme, the organization might insist that each employee manage his or her own mail server[2]. This gives the organization the least control over its email. A more middle-of-the-road scenario might be each department managing its own email server. The centralized server offers the least flexibility because no departments have any control over their email. The most flexible architecture is to have all users running their own mail servers — everybody can do as he or she pleases. The structure with some flexibility has each department manage its own mail server — then each department is free to do what it wants. Different management structures offer different levels of control and flexibility.

Figure 4.3(a) and (b) illustrates two of these management architectures. The centralized structure in (a) allows the organization to change as a whole. Changes affect all departments, so enhancements are hard to plan

---

[2] This extreme is unlikely, but still possible.

and implement. In (b), different departments manage their own email servers. This might be very practical — Accounting and Finance might be best served with LAN-based systems, such as Lotus Notes or Microsoft Exchange, while Engineering might want the flexibility of Unix architecture with its well-known Pine or Emacs email readers. This distributed structure gives users the flexibility to choose their email servers and readers. It allows each department to decide when and how to upgrade its email services, thus allowing more experimentation than the centralized, single-mail-server architecture. In this case, experimentation occurs at the departmental level — a useful structure when there is uncertainty because it allows each department to find what best meets its needs.

In some organizations there is a combination of centralized and distributed structure. Harvard is one example of this because it does manage a centralized email server that is available to central administration. The separate schools, such as Law and Business, also run their own mail servers, as do some individual departments within the schools. Computer Science is one example. Harvard also allows an individual to run a mail server at his or her desktop. My colleague, Scott Bradner, does this — he likes the control and flexibility it gives him. This combination approach is likely to work well in many organizations because it allows flexibility where needed and efficiency when service needs are well defined.

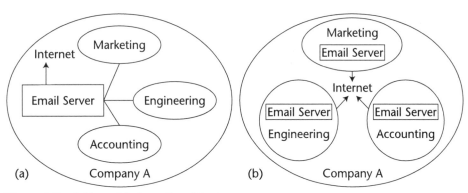

Figure 4.3 Self-provided email architecture.

Figure 4.4 illustrates how to apply the framework in Figure 4.1 to the particular example of email. At the top layer, the outsourcing decision is made — if market uncertainty is low enough, then outsourcing is likely to work well and is a good choice. Assuming outsourcing, the next decision is the choice of Web-based or ISP email. This decision determines the next choices. If ISP email is chosen because of its more distributed management style, then there is a protocol choice to make — POP or IMAP. Post Office Protocol (more distributed) requires that the user manage his or her own email messages. Internet Mail Application Protocol (IMAP) is different; it gives users a choice, and it allows both centralized and distributed management of the messages at the user's discretion. This is the lowest level of the ISP branch of the hierarchical structure. If the self-managed email server is selected by branching right at the top layer because of high market uncertainty, then the next layer is the architecture decision of how to structure the email service for the organization. Both distributed and centralized architectures are possible — high market uncertainty within the organization would suggest that the right-hand tree limb with its distributed architecture is best. This example illustrates how this framework is a general tool to apply to designing management infrastructure. It helps compare the value of flexibility with distributed management to the business and technical advantages of the centralized management structure with market uncertainty factored into the decision.

These examples show a variety of ways to provide email to the organization. Each strategy allows a different level of control and ability to experiment. Outsourcing implies that the user has no ability to experiment. Providing your own service implies that you have the ability to experiment to meet uncertain needs, but the level of this ability to experiment depends on how centralized the internal system is. Today, the decision is complex because with voice/data convergence the reality of unified messaging is closer, yet still very uncertain. I for one am not sure how my voice and email should best be integrated. I need to see several ideas, and then I might need to try them out. The point is that I don't know what I want and that I need to see the choices to decide — this is the definition of market uncertainty. Its link to the value of experimentation is important because it helps us make decisions leading to good management structure for the given organization.

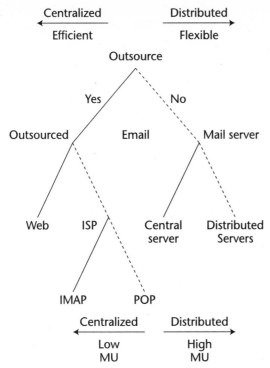

Figure 4.4 Hierarchical structure of management architecture of email.

Voice

Voice services have a long, rich history (see Chapter 9), and today a company has many choices to provide these services. As in the email example, voice services may be outsourced or provided in-house. Similar to email, once the decision to outsource or self-manage is made, there are additional choices. Both basic (such as call forwarding and speed dialing) and advanced voice services (such as voice mail) are available as an outsourced service called Centrex. It provides the same services as a Private Branch Exchange (PBX), but with Centrex the service provider manages the service.

If the organization decides outsourcing is not the best answer because its success depends on cutting-edge voice features such as computer/telephone interfacing for advanced customer services, then owning and managing a PBX gives the company more control and flexibility in providing these services. This approach to managing voice services requires decisions about the architecture of the PBX design. Should a single large PBX be used, or is a distributed group of PBXs a better answer? It turns out that all these ideas work best for some users depending on the market uncertainty for that user group. The case study in Chapter 9 shows that when market uncertainty decreases, the percentage of users satisfied by systems with more centralized management architectures increases.

For many users, outsourcing voice services makes a lot of sense. It allows a company to focus on its core business — not dealing with a PBX and its management. The telephone company is good at providing robust and reliable voice services, and Centrex services have matured and become more advanced over the last 15 years so that they meet many users' needs well and are priced competitively to PBX ownership. Figure 4.5 illustrates this centralized architecture. The nature of telephone companies allows them to provide service to distributed locations; it even allows workers at home to have similar capabilities to the office environment. Many large distributed organizations, such as universities, find that Centrex is a good solution. Centrex is popular because it meets the needs of many users. For many users these voice services are a commodity. These users don't need flexibility because they care only about the stability offered by outsourcing voice services.

There are many companies whose needs are not met by Centrex because they demand control, flexibility, or the most advanced features in order to have a competitive advantage. Owning and managing a PBX does allow the organization to control upgrades, providing new services when they choose. Owning and managing your own PBX gives users many more choices because there are more PBX vendors than local access providers. Even though most PBXs offer similar features such as call forwarding and transfer, caller ID, and speed dialing, the high-end systems do differentiate themselves by offering advanced features, such as Automatic Call Routing (ACD), and a computer/telephony interface. There is more experimentation with PBXs than Centrex services because of the larger number of PBX vendors and the nature of PBX systems as compared to CO switches. In general, it is less expensive for PBX vendors to experiment with new features, which implies that you expect the most advanced services to first appear on PBXs and then migrate to Centrex services, as discussed in Chapter 9. Some users will never be happy with Centrex because it lacks the most advanced features and its centralized control is constraining to many.

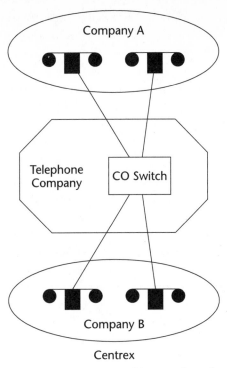

Figure 4.5   Centrex architecture for voice services.

Depending on the organization's needs and the market uncertainty, a PBX might be a good idea, but this decision is just the tip of the iceberg. The harder part of managing your voice services is deciding how to structure them. Should they be centrally managed by the organization, or should each department assume responsibility for management of its own PBX? It depends on the organization because each has advantages and disadvantages. Figure 4.6 shows two architectures for a PBX: (a) is a centralized model while (b) has a distributed structure. The more centralized the management is, the more efficient the management of the PBX will be, but it will also be less flexible. The needs of all departments might be better met by giving each department a choice of PBX vendor.

The flexibility of a distributed PBX design might be justified in organizations that have departments with different needs. The more distributed architecture shown in Figure 4.6(b) allows each department to select a PBX that best meets its needs. Giving departments this flexibility to choose the system is the best strategy for uncertain environments. Staying on the cutting edge with customer service might mean frequent upgrades to a PBX; however, other parts of the organization might be less demanding and are satisfied with less frequent upgrades because they need only basic voice services with simple voice email features. This distributed model allows each department to experiment and meet its needs — something not possible with the centralized architecture.

This centralized or distributed structure is not an all-or-nothing proposition. Organizations might have both structures. A large centralized PBX might serve most departments, but some areas, such as customer service and order processing, may have a more specialized system for their more demanding tasks. With the integration of voice and data and the growth in Web-based services, having the flexibility to experiment might be a critical success factor. Now, with Session Initiation Protocol (SIP) a part of Windows XP, voice-enabled Web services are likely to become ubiquitous. The uncertainty in this area is high because nobody knows what these services will be or how they will integrate with the current voice services within the organization. Giving departments that need to experiment with advanced technology (such as SIP) the ability to do so is critical to keeping up with these new voice-enabled Web-based applications. Sometimes multiple management structures within a single organization will best meet users' needs.

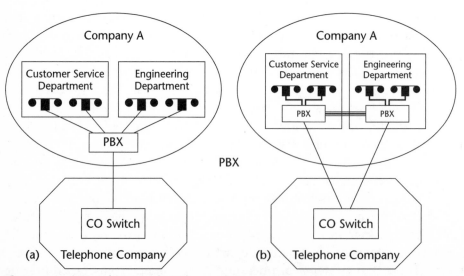

Figure 4.6   PBX architecture for voice services.

The choices a manager needs to make to provide voice services are illustrated in Figure 4.7, showing the hierarchical structure of these decisions. As in the email example, the first choice to make is whether to outsource voice services with Centrex. Low market uncertainty implies that Centrex might work well. On the other hand, if PBX ownership and management is picked, then this figure shows that the next-level decision is about the architecture of the PBX. The choice is between a distributed and centralized structure, with market uncertainty one of the important factors influencing the decision. Similar to email services, voice services fit nicely into this hierarchical framework.

The choice of management structure for voice services is complex because many factors affect the decision. In some cases, market uncertainty drives the choice of architecture because the flexibility to experiment becomes the most valuable attribute of the management structure if uncertainty is high enough. Distributed architecture is good in uncertain environments, as shown when comparing distributed and centralized PBX structures. If you don't know what each department wants, how can you provide a centralized solution? The only way to meet uncertain department needs is to allow experimentation to figure out what these needs are. Once these needs are determined, it might make sense to manage them centrally.

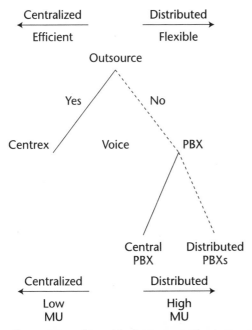

Figure 4.7 Hierarchical structure of management architecture of voice services.

Information

The Web is the best way to distribute information for many organizations and individuals because it gives users an easy-to-use platform that reaches a broad and growing international audience. It is a given in today's world that Web-based services enhance access to important information. It's not clear, though, how these services are best provided — should they be outsourced or built by the organization? As with email and voice services, there are many layers of decisions to make, and these decisions depend on the market uncertainty. At the top layer is the decision to outsource or manage the services in-house. Outsourcing makes the most sense when market uncertainty is low because it is easy for the service provider to meet the customers' needs. If it is outsourced, then you need to decide who should manage the content of the Web site, which again depends on the market uncertainty. If you decide to manage your own Web server (higher market uncertainty), then what should the architecture of the infrastructure used to provide Web services be? Will a large centralized server (better in low market uncertainty) or many smaller, distributed servers working in parallel better meet your needs? Again, there is certainty about the need for this service, but there is uncertainty about the best way to provide this service for a particular organization.

For many organizations, outsourcing their informational needs is the right answer. There are many Web hosting companies that will help develop and manage the service. This works well with smaller companies that don't have the technical skill to manage their own Web site. Consider a company called Irwin R. Rose & Co, Inc., which is an Indianapolis, Indiana based owner and operator of apartment communities doing business in five states. The president, Doug Rose, is not a technologist, nor does he have any on his staff. There is no reason to have one because his business is renting apartments, not building Web sites. However, Doug Rose knows that he must have an Internet presence because more renters are finding what they need over the Internet. Doug's solution was to outsource the development and management of his Web site. Outsourcing is a good fit for organizations with critical success factors not directly related to their Web presence.

Figure 4.8 shows the architecture of the outsourced solution. This is a carefree environment for the user because management of the Web server is by the service provider. When outsourcing Web services, the user has varying degrees of control. In some cases only the service provider can alter Web content, while in other cases the user has Web-based control to manage the content, but on the service provider's server. Yahoo! Storefront is a service like this. Its roots are in a company called ViaWeb, which was the first service provider to build a system that let its customers create and manage an online store. ViaWeb managed the servers and payment, while the store managed its content. This was a good match for what the market was looking for. No matter what degree of control the user has over content, the outsourced solution means that the user does not need to manage the hardware and software of the server. The degree of market uncertainty helps determine the best structure for outsourcing and helps the user decide how much content, if any, he or she wants to manage.

Yahoo!'s main business is its Web site. It would never consider outsourcing its development or management. Because adding new services is Yahoo!'s lifeblood, control of its Web-based services is a critical success factor and therefore must be under Yahoo!'s control. Organizations such as Yahoo! need more control and flexibility than any outsourced solution is capable of providing. Being able to decide when to upgrade to new versions of HTML or HTTP is critical to Yahoo!'s success, and putting these decisions in the hands of others would be unwise for any organization with a competitive advantage that depends on the creativity and robustness of its Web site.

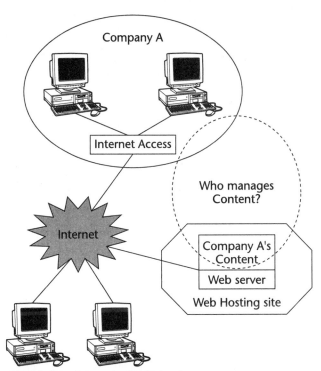

Figure 4.8 Outsourced information server.

Figures 4.9 and 4.10 illustrate the centralized and distributed structure possible with a Web-server. The centralized model depicted in Figure 4.9 consists of large servers, such as Sun's Fire 15K, 12K, and Enterprise 10000 servers or IBM's zSeries 900. Servers like this can cost over a million dollars. These systems are robust, and the vendors provide good support. This efficiency comes at a high price — it's not easy to experiment with new services on large centralized servers. This difficulty of experimentation is due to the large number of users depending on each server. Managers of these large systems exercise extreme care when upgrading applications. The more distributed architecture in Figure 4.10 is far more flexible in terms of experimentation. In this distributed structure, Web requests are routed to one of many Web servers. No single Web server is critically important. With this type of architecture, it is easy to change the software of one server. If the new software is bad, it is not a catastrophic event, as it would be with a large centralized server, because a smaller, more isolated group of users is affected. The centralized architecture in Figure 4.9 works best in low market uncertainty, and the distributed structure in Figure 4.10 meets the needs of users better when market uncertainty is high because it allows more experimentation.

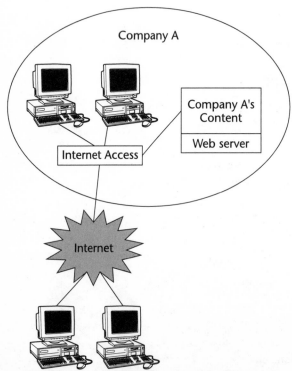

Figure 4.9   Centralized in-house information server.

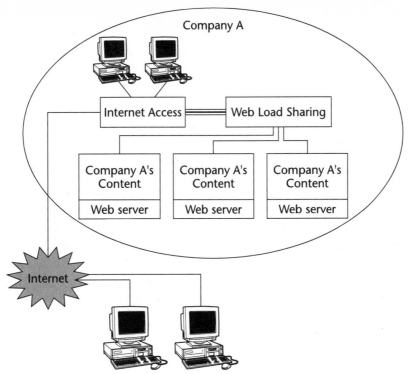

Figure 4.10 Distributed in-house information server.

How should organizations structure their internal Web server infrastructure? Reading the trade rags brings only confusion. Sun and IBM suggest that expensive large servers are the best architecture, but other vendors recommend using a greater number of less expensive servers based on the Intel architecture. Both groups present arguments that are logical. Large centralized servers are indeed easier and less expensive to manage and have extremely high reliability. Server farms of smaller systems, while harder to manage, are easy to expand incrementally, have good reliability because of the redundancy of having many servers, and allow easy experimentation with new services. Yahoo!'s decision to build a distributed infrastructure fits well with the high level of market uncertainty surrounding the Web-based applications it provides.

Yahoo!'s architecture of a server farm of Intel-based PCs running a type of Unix allows Yahoo! to experiment with new services very effectively. By installing a new service or changing an existing one on a single server, only a small fraction of users will see this new or changed service. If things don't work, it's easy to back out. On the other hand, if the experiment is a success, it's easy to phase in the new service. This allows Yahoo! to experiment with many new services and pick the best ones. It provides a way to gauge results because the new system can be compared to the old system in a controlled way. There are many trade-offs to consider when looking at server architecture, including flexibility, ease of experimentation, and business and technical advantages. In Yahoo!'s case, the degree of market uncertainty, coupled with the company's needs, implies that the distributed architecture makes sense.

The choices managers make to provide Web-based informational services are illustrated in Figure 4.11, which shows the hierarchical structure of this decision. Similar to email and voice, the first choice to make is whether to outsource managing the Web server. Low market uncertainty implies that outsourcing this function might work well. Management must decide who manages the content of the Web site: the service provider or the user. On the other hand, if Web-server ownership and management is picked because of higher market uncertainty, then the next-level decision is about the architecture of the server infrastructure — should it be distributed or centralized? As with email services and voice services, Web-based informational services fit nicely into this hierarchical framework where market uncertainty helps decide whether a distributed or centralized management structure will create the most value.

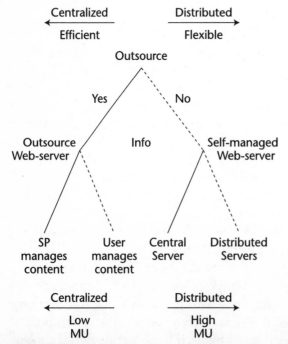

Figure 4.11   Hierarchical structure of management architecture of informational services.

There are good arguments on both sides of this issue about how to structure a large-scale Web server installation, making the choice of architecture difficult. The answer again boils down to market uncertainty and the ability to meet user needs. High uncertainty implies that the value of experimentation from the distributed model exceeds the benefits of consolidating many smaller servers into a centralized unit. When market uncertainty is low, then the ability to experiment is of less value because experiments have similar results. Low market uncertainty means you don't need to experiment to see if particular users will accept or reject the service — you know the result of the experiment without performing it. Market uncertainty is a key factor in decisions pertaining to the choice of management structure.

Conclusion

This chapter discussed the management structure of network-based services in more detail. It explored why experimentation is so hard with a centralized structure and why distributed management architecture promotes innovation through experimentation. It examined the hierarchical structure of management architecture and how market uncertainty factors into the decision about management structure at each level. It discussed three different network-based services (email, voice, and informational) and illustrated different management structures for each of them and how they coexist. The discussion about email examined outsourcing compared to internal email management. It showed the flexibility of the distributed management solution compared to the easy management of the centralized outsourced model. Next, voice services were discussed in the context of whether an organization should outsource its voice services with Centrex or buy and manage a PBX. It also looked at the advantages and disadvantages of a centralized versus a distributed architecture of a PBX. Last, this chapter discussed how to provide information over the Web. Outsourcing was compared to managing the Web server, and different Web server architectures were examined. The flexibility of the server farm model was compared to the efficiency of a large centralized server. These three examples illustrate how different management structures can coexist and how each structure meets the needs of some users best. It is the market uncertainty of the particular user that is important in the choice of management architecture.

The next chapter examines the theory of options and its extension to real options, which form the basis of the ideas in this book. It takes the intuitive approach by leaving out the complex mathematics behind the theory and explains in practical terms how and why understanding options creates value from uncertainty. It is important because it is the foundation behind the theory in this book about why high market uncertainty requires a management structure that allows easy experimentation. Understanding how to think with an options point of view enables managers and investors to gain value from uncertainty — a powerful tool in today's crazy world. Options theory illustrates the value of managerial flexibility by showing the value of allowing managers to alter course as current conditions change. Understanding options thinking is a critical advantage in today's uncertain world.

Intuitive View of Options Theory

Investors and managers that understand how to think in terms of options are at a competitive advantage in today's uncertain world. Options were invented to help manage uncertainty in financial markets. The options way of thinking is critical to the modern manager as a strategy to profit from uncertainty. The greater the uncertainty, the greater the potential value of thinking with this options framework. The options point of view means a focus on flexibility and the ability to evaluate the current situation and alter the strategic path to better match the contemporary environment. The manager or investor thinking with the options mind set is not afraid of uncertainty because he or she understands how to craft opportunities that have increased value because of the uncertainty. Real options are the expansion of this theory to help manage non-financial assets in uncertain environments. Learning how to create value from uncertainty is a good reason why understanding options is important to modern managers and investors.

This chapter examines the options theory and real options at an intuitive level. It illustrates how this theory helps managers make choices that maximize value in uncertain conditions. It shows the link between uncertainty and value. Several applications of real options are presented: first the value of a staged investment strategy when building an Information Technology (IT) infrastructure and, second, the value of modularity in computer systems. This argument about the value of modularity is the underlying principle of the theories presented in this book. Although not complete, this chapter provides a flavor of what options theory is and why it is important to managers and investors.

Options Theory

The theory of modern finance seeks to allocate and deploy resources across time in an uncertain environment [1]. Traditional finance models such as the Capital Asset Pricing Model [2] show how to optimally deploy assets by determining the optimal path over time given the initial conditions. These older methods, though, do not adequately capture the value of management having the flexibility to change strategies in response to unfolding events. The theory of contingency finance forms the framework of options theory, which addresses this problem by allowing management to revise decisions continuously over time [1]. This new view allows valuation when agents (specifically managers) are able to make choices as events unfold. It shows how uncertainty increases the value of a flexible asset deployment strategy.

The theory of options has proven useful for managing financial risk in uncertain environments. To see how options can limit risk, consider the classic call option: It gives the right, but not the obligation, to buy a security at a fixed date in the future, with the price determined in the past. Buying a call option is the equivalent of betting that the underlying security will rise in value more than the price of acquiring the option. The option limits the downside risk but not the upside gain, thus providing a non-linear payback, unlike owning the security. This implies that options provide increasing value as the uncertainty of the investment grows (that is, as variance in the distribution describing the value of the security increases and the mean is preserved).

Options are a powerful methodology of risk reduction in an uncertain market. They allow limiting loss without capping the potential gain. Figure 5.1 shows graphically what the value of an option is. The non-linear payback of the option is the solid line, while the linear payoff of owning the stock is the dashed line. The option holder is able to look at the price of the security when the option is due and decide whether to exercise the option to buy the stock. This protects the option holder by limiting the loss to the cost of acquiring the option, no matter how low the stock price falls. Some risk-averse investors prefer this type of non-linear payback that caps the downside risk, but leaves the upside gain unaltered.

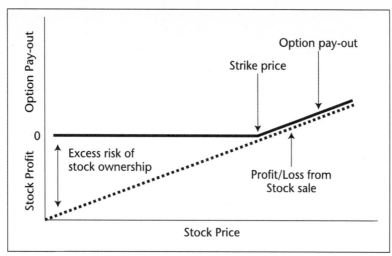

Figure 5.1 Value of an option.

The following numerical example further illustrates this concept. Suppose stock xyzzy is selling at $100.00 on January 1, 2002. The classic call option is to pay a fixed amount per share (say $10) for the option to buy shares of xyzzy on February 1, 2002 at $100.00 per share. Suppose you buy this option on 100 shares of xyzzy for $1,000. If xyzzy rises to $200.00 per share, your profit is the gain in the stock minus the cost of the option: 100*$100 – $1,000 = $9,000.00. If xyzzy does not rise above $110, the option is worthless, and you lose the full cost of the option ($1,000). This is the most you can lose, even if the stock falls to 0. Compare this to owning the security. Buying 100 shares of xyzzy costs $10,000 on January 1, 2002. If its value increases to $200 a share, and you sell it on February 1, $10,000 is the profit. If the price does not change, you lose nothing. If xyzzy files for bankruptcy and the stock drops to zero, you lose the full $10,000. The difference is dramatic between owning the stock and owning an option to buy the security. The options cost $1,000 but protects from a possible loss of $10,000. If the stock takes off, the upside gain is offset only by the fixed $1,000 cost of the option. This example shows how options limit the risk without capping the gain of an investment.

The historical variability of the stock price, not the average price of the security, determines the value of the option. This means the variance of the distribution describing the option value is independent of the average value of the asset. This is intuitive if you consider the following example:

Stock xyzzy is very stable, and very expensive, at $1,000.00 a share. For many years its value has remained constant so its average price is $1,000. Any option on this stock is worth little because the uncertainty of what the price of the stock will be is so low — it most likely will be very close to $1,000. On the other hand, consider the stock from zzzz, which has been bouncing around for two years. The low was $10, with a high of $200, and an average price of $20. The option to buy shares of zzzz is valuable because there is a chance of the stock rising again as it did before. This illustrates that the value of an option is based on the variability of the price, not its average value.

Understanding the complex mathematics underlying the theory of options is unnecessary to gain a good intuitive understanding of how this theory can help convert uncertainty into profit. One key to understanding this options point of view is realizing that the expected value of a correctly structured investment portfolio can increase as uncertainty grows. Options are about the relationship between uncertainty, flexibility, and choice. The idea is that increased uncertainty amplifies the value of flexibility and choice when the correct strategy is crafted.

In technical terms, options help value choice in capital markets.[1] The capital market aspect means the players in the market determine the value of each choice, and this value is random. If the values are known, then dynamic programming is the correct valuation technique to use. In simple terms, the complex partial differential equations explaining the theory of options are a mathematical model describing this value of flexibility. It illustrates how more uncertainty about the value of choices in the market increases this value of having choice.

This book is about the value of users having choices, and options are a proven methodology to value this choice. Think about two situations: a user having only one choice for a service or a user having many choices. If there is uncertainty about what the user will like and how much the user will value the service, then the value of the service is random. Options thinking illustrates that this value of giving users choices is the difference between the expected value of each choice, subtracted from the expected value of the best of all the choices. This makes intuitive sense because the expected value of the best choice is the value a user gets if he or she has a choice to pick what they prefer, and the expected value of any particular service is the average value for the service if this user only has one choice. This book and the options framework illustrate how the value of allowing users to pick from among many choices grows as market uncertainty increases.

---

[1] This understanding comes from a conversation I had with Carliss Baldwin at Harvard Business School while working on my thesis.

Real Options

This theory of options is extendable to options on real (non-financial) assets [3]. Real options provide a structure linking strategic planning and financial strategy. Similar to financial options, real options limit the downside risk of an investment decision without limiting the upside potential. In many cases, this approach shows a greater potential expected value than the standard discounted cash flow analysis performed in most corporate environments. This theory is useful in examining a plethora of real-world situations such as staged investment in IT infrastructure [4], oil field expansion, developing a drug [3], the value of modularity in designing computer systems [5], and the value of modularity and staged development of IT standards [6]. This expanded theory of options into real-world assets is proving its value because of the wide range of applications for which this theory has proven useful.

Staged IT Infrastructure Development

There are several ways to build a technology infrastructure. Investment and implementation can be in one or several stages. Staging the investment required to build large Information Technology and telecommunications systems provides an option at each stage of the investment. The option is whether to continue the investment, and it is based on the most current information available about the uncertain market, economy, and technical attributes of the project. Starting out small and evolving the project at various stages allows making more focused and relevant decisions, which in turn increase the expected value of a staged implementation over that of the single-stage scheme if the market uncertainty is great enough to offset the extra cost of staged development.

Investing in IT infrastructure in stages offers management more flexibility, as shown in Figure 5.2. It shows a theoretical IT project that has two stages — phase 0, the beginning, and phase 1, which is after the decision point to evaluate phase 0. Traditional financial methods such as Discounted Cash Flow (DFC) analysis do not give management the flexibility to alter the course of a project. Options theory allows the project to be evaluated at different stages with the investment strategy altered according to the current environment (which may be very different from the environment predicted when the IT project began). Giving management the flexibility to change course during the project is the hallmark of the options approach. It provides a more realistic value for the IT project because it more accurately reflects the choices that management really has. Projects do get cancelled all the time, yet traditional financial methods don't factor in this basic fact. Options provide a better way to view the costs of IT projects.

The advantage of the staged approach illustrated in Figure 5.2 is that it allows more possible outcomes of the IT infrastructure being built. With a single phased implementation there are only two outcomes: not investing in the project or investing in and implementing the entire project. These are the top and bottom paths (bold lines). With a phased implementation there are several other possible outcomes: investment in or abandonment of the project after the end of phase 0. If investment is the choice management makes, it then has other choices: It may invest at the budgeted amount, invest less than planned, or increase the previously planned investment. If the project is less successful than estimated but still worthwhile, then it might make sense to cut the planned investment. On the other hand, a successful project might imply that scaling up the project makes the most economic sense. This staged strategy shows a higher value because the project manager is able to determine the best investment option at the end of phase 0 depending on the success of the project and the current conditions. When uncertainty is high, the value of having these choices along the development path increases.

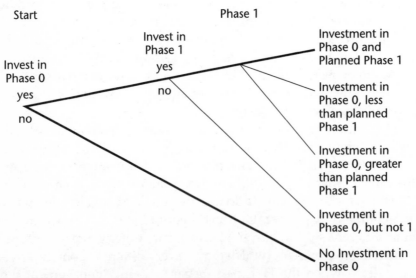

Figure 5.2 Staged investments in IT infrastructure.

Value of Modularity in Computer Systems

In "Design Rules" [5], Baldwin and Clark apply a real options theory to study modularity in the computer industry. They show how modularity in computer systems (like the IBM 360) changed the industry tremendously. Modularly designed computers consist of components that have defined interfaces. Because each component conforms to its interface rules, modules that follow the defined interface are interchangeable. In contrast, an interconnected system has no swappable components because only a single massive component exists. Baldwin's work shows how modularity increases value and how increasing technology uncertainty about the value of the modules increases this value of modularity.

To see how a modular design provides value, consider the evolution of a computer system without a module design. Figure 5.3 illustrates such a system: It consists of four main functions — storage, memory, I/O, and the CPU. Suppose that this computer is being redesigned and both the memory and the CPU are changed. Now, assume that this redesigned CPU worked well and increased the value of the total system by +1; however, the new memory design did not work as expected. It decreased the value of the total system by –2. When redesigning a computer that has its functional pieces interconnected, the new artifact provides a single choice; the new system performs, as a whole, better, worse, or the same than its predecessor does, and you can take it or leave it. In this case, the new system has a value less than the original system; it is the failed memory experiment that drags down the total value of the system. The interconnected architecture of this computer does not allow the choice of using only the improved CPU, without the inferior new memory design.

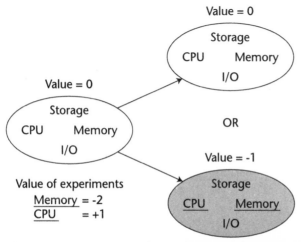

Figure 5.3 Interconnected system.

In contrast to the interconnected design previously mentioned is the modular computer system illustrated in Figure 5.4. If the same redesign is attempted with this modular architecture, there are more choices for the new system when compared to the interconnected system, a similar effect to having more possible outcomes with staged development, as discussed previously. The system with the most value uses only the new CPU, keeping the older, but better-performing memory design. This new system has a value of +1, higher than the interconnected system allows. As with all options-like situations, increased uncertainty implies increased value of the modularization. The modularity allows the system designer to pick and choose the components of the new system, thus maximizing the value. Uncertainty increases this value because as it grows it increases the potential of a really great choice emerging. The modular design increases value by providing a portfolio of options rather than the less valuable option on a portfolio.

Baldwin and Clark [5] computed the value of modularity. Let V_1 be the value of a complex system built as a single module, and let Vj be the value of the same system with j modules. If the cost of modularity is ignored, then the value of dividing a complex system into j components is: $Vj = j^{1/2}V_1$. That is, the modularized system exceeds the value of the interconnected design by the square root of the number of modules. This value does not depend on the variance of the distribution because for each module there is only one choice — keep the old or use the new — and there is only one choice for the new module. This is intuitive; if you take a single sample from a random distribution, the expected value is not dependent on the variance. If there are many choices for the new module, then the variance of the distribution is important to the expected value of the best of many choices.

This modularization allows the designers to experiment on modules that have the most potential for altering the value of the system. Each experiment is one design of the module. Performing many experiments on the components most critical to overall system performance maximizes the overall value. Because of the modular design, the designer now has the option of picking the best outcome from many trials. For example, suppose the designers of a new computer system need to increase the rate at which a CPU module processes instructions, as illustrated in Figure 5.5. (This is similar to the idea behind Figure 6.1 in the next chapter.) It shows that three attempts are made to improve the CPU; the worst experiment lowers the value of the total system by –2, and the best new design increases the total system value by +2. By attempting several technically risky new technologies for a CPU, the designer can improve the odds of reaching the goal of faster instruction execution. In this case, the best CPU increases the value of the total system by +2. The modularity allows system designers to focus on components that have the greatest potential to increase the value of the whole system.

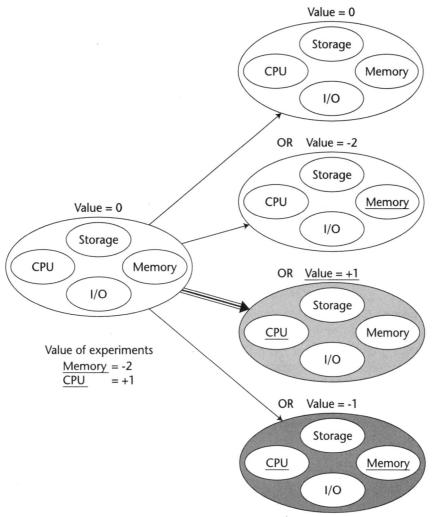

Figure 5.4   Value of modular system.

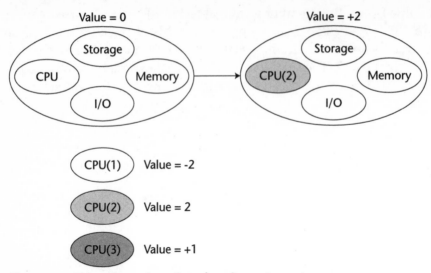

Figure 5.5 Value of experimenting where it counts most.

This modularity allows picking the best from many experiments on a module, which is illustrated in Figure 5.5. In the preceding example there were three experiments for the CPU, and the best of these experiments yielded +2. Assuming that the experiments on the CPU are normally distributed, then the values of the experiments are expected to look like Figures 6.1 and 6.2 in the next chapter. Each experiment creates a choice to use a particular module from many different choices, which is the same as picking the best of many experiments. As the uncertainty about the technology used in the module grows, so does the value of having a choice from many modules. If uncertainty is high, and if you have enough experiments, it becomes likely that an exceptional (as defined as standard deviations of +3 or more from the mean) new module is found.

This value of picking the best module from many choices is similar to the value gained from a distributed management structure because the experimentation it allows enables users to have many choices of network-based services. Interconnected systems make experimentation difficult much the same way that central management hampers the ability to experiment. Modularity gives designers choices in the same way that distributed management gives users choices. This model of the value of modularity forms the framework for this book because of this mapping from the value of modularity to the value of distributed management. Both have the most value in high uncertainty because of the increased value of experimentation enabled by module design and distributed management structure.

Value of Staged Development and Modularity in IT Standards

In these papers Gaynor and Brander [6][7] propose a model of technology standardization based on modular standards and staged development. This theory is based on the effect of market uncertainty on the value of modularity. A real options model quantifies this value of modularity in standards, illustrating that a rational way to standardize some IT technology in uncertain markets is with modular design, end-2-end structure, and proper staging of the standard. Highly modular standards provide a higher option value because of the ability to select the best modules to change at a fine granularity; the staged development promotes the best use of the modularity.

These papers argue that standard development methodologies promoting a broad range of experimentation, combined with market selection, will better serve end users by involving them in the standardization process. In high market uncertainty, promoting experimentation with new proposed standards and standardizing the technology adopted by most of the community decrease the risk of an unaccepted standard. Design principles, such as the end-2-end argument, that push intelligence to the network's edge help promote this type of experimentation because they broaden the range of participants able to innovate by allowing users to create services.

These papers propose a prescriptive model of technology standardization under uncertainty. This model of standardization is simple and intuitive: Start with simple standards structured in a modular layered architecture, then let the standards evolve, with the market acting as the selection mechanism. This model illustrates how modularity in standards creates value and how market uncertainty increases this value. This work explains how to apply this framework to the development of communication protocol standards. It illustrates that modularity (up to a point) intrinsically creates greater value than an interconnected design. It argues that standards that allow applications to have end-2-end structure, where the network provides only basic transport services by pushing applications to intelligent end points, create an environment conducive to experimentation. For example, the network layer in a network protocol stack (IP) should contain protocols providing only the most basic services, while the application layer should contain protocols offering the most diversity in terms of the services offered. These papers discuss the value created by applying the methodology of introducing protocol suites (and protocols) with a few simple modules and evolving the standards by creating new protocol modules or altering existing ones. The theory shows that the evolutionary staged approach to development of entire protocol stacks, and protocols within each layer, maximizes the expected value of the standard.

Conclusion

This chapter is an intuitive approach to an unintuitive idea — it demonstrates how uncertainty creates value. Understanding options, and real options at an intuitive level, illustrates how to craft a strategy that creates greater value with increased market uncertainty. Strategies based on options thinking value flexibility because of its ability to capture value from the unexpected. Looking at examples of applications of real options to an IT infrastructure and modularity provides an intuitive feeling for how to structure strategy with this options mindset. The most important part of this chapter is the last part linking the value of modularity to the value of distributed management. This mapping is a key idea behind the value of giving users choices. Intuitively, it makes sense that the value of having many choices increases when the values of the choices vary greatly.

The next chapter discusses market uncertainty and how it changes the value of experimentation. The chapter looks at other research about market uncertainty. It discusses how to measure it and gives several examples. Next, the effect of market uncertainty on the value of experimentation is demonstrated. It illustrates how high market uncertainty enhances the value and how low market uncertainty suppresses this value of experimentation. Linking market uncertainty to the value of experimentation is the framework for this theory.

Market Uncertainty

Understanding what market uncertainty is, and how best to manage it to maximize gain, is essential to success when investing or managing in today's world. Even before the September 11 tragedy, uncertainty in network-based services was high. Now, smaller budgets and heightened concern for security make it even higher. Thus far, it has been very hard to predict which services and applications users would embrace. Ideas such as online grocery shopping (WebVan, PeaPod, and so on) attracted over a billion dollars from investors, and they are now worth very little — what seemed like a good idea was not. Yet, other ventures, such as online auctions (eBay), have proven valuable to investors and users. Other services, such as online map/direction services (MapQuest), have proven valuable to users but have not found a business model to ensure their continued existence. Market uncertainty is high regarding network-based services. Investors and managers who understand how to use uncertainty to their advantage will have a strategic competitive advantage over their peers.

Understanding how market uncertainty changes the value of experimentation is an important strategic advantage. When market uncertainty is high, then being lucky with a correct guess about the market is likely to produce more revenue that being right in markets with low uncertainty. Being right when market uncertainty is great implies that competition will be feature-based, and being right means you picked the best features.

When market uncertainty is low, most vendors and service providers are doing the same thing, which causes the competition to be priced-based, rather than feature-based. It is harder to win big in a commodity market because users have many choices that meet their needs. This fact forces vendors and service providers to lower usage fees to gain a bigger share of the market. With high market uncertainty, the odds are against you, but being right produces great wealth; when market uncertainty is low, your chance of meeting the market is great but unlikely to yield large profits because of price sensitivity.

This chapter defines market uncertainty, how to measure it, and its link to the value of experimentation. It starts with a discussion about market uncertainty in a general sense by looking back at what others have discovered. Then, it discusses several methods to measure market uncertainty. These measurement techniques include both established methodologies and ideas presented in my thesis. The chapter ends by linking market uncertainty to the value of experimentation.

What Is Market Uncertainty?

Market uncertainty is the inability of vendors and service providers to predict what users will like. Market uncertainty is not new; it has existed for many years. With today's fast-changing telecommunication and computer technology, however, it has reached new heights. The uncertainty exists partly because users often do not know what they want until they see it. This means that service providers cannot ask users what they want. The only way to meet uncertain markets is by trying different ideas and hoping to find at least one that will work.

Users often do not know what they want from new technologies because they don't know enough about them to understand the possibilities. As previous research from Clark [1] shows, when users are first introduced to new technology, they tend to view it in the context of the older technology being replaced. Users' expectations evolve along with the technology as they become more educated about the technology and what it enables. Clark noted that when the first automobiles were built, users viewed them in the context of a horse-drawn carriage (hence the name "horse-less carriage"). Only later, as users began to understand the range of possibilities, did attributes such as reliability, comfort, and safety become important. This illustrates how users' preferences evolve as they understand more about a particular technology.

As Clark points out, market uncertainty is hierarchical in nature. Consider the migration from horse-drawn carriages to automobiles. At first, the uncertainty existed with high-level design questions, such as what type of power plant is best to replace the horse. Next, decisions such as the best steering, brakes, and tires became important. It became obvious that the tiller design used with the previous technology did not meet the needs of the new technology. In today's car, there is little uncertainty about the power plant[1], steering, or braking. Consumers today are more concerned with safety, efficiency, and performance, not with the basics of how the car works. It is this combination of new technology and users' perceptions of their evolving needs that creates market uncertainty.

A similar phenomenon is occurring with the Internet and the Web. The diversity of Web-based applications is beyond what pundits ever imagined. Nobody predicted in the early 90s what the Web is today or its impact on society. In 10 years, the Web has emerged as a requirement for modern commerce. The Web is the standard way to share information, both within a company (intranets) and to the outside world (extranets). Web-based services from banking, to shopping, to travel and even sex have become the norm for many. The Web has enabled customers to help themselves to services and information without having to depend on customer help lines. The Internet today is far different from the predictions of early 1990. This shows the high level of market uncertainty that exists in network-based services and the way users' preferences evolve with the technology.

Understanding market uncertainty is important for product design and development. The Internet has changed how applications are designed and built because of the high uncertainty of the Internet environment. Think about development of the first breed of Web browsers. When Netscape started its development process [2] there was extreme uncertainty. Users had no idea what they would do with browsers, and vendors had no idea what services would become popular. Understanding the high market uncertainty, Netscape altered the traditional software development process to allow for extraordinary levels of early feedback from users. It also changed its software development processes, allowing incorporation of this feedback into the product design at advanced stages in the development process, when traditional software engineering methodologies would not allow changes. Netscape succeeded in its browser development because it understood how to take advantage of the extreme market uncertainty.

How to Measure Market Uncertainty

It is important to be able to measure what you are trying to manage; thus, measuring market uncertainty is important. Knowing if market uncertainty

---

[1] This is changing because of environmental concerns. The efficiency and environmental friendliness of the combustion engine is being questioned, but it has been a stable, dominant technology for over 50 years.

is high, medium, or low is important to shaping the management policy of network-based services. Although the ability to measure is critical, the precision of the measurement is not. In fact, precise measurements of market uncertainty are not possible. Fortunately, it is possible and sufficient to estimate market uncertainty as low, medium, or high.

I try to use techniques for estimating market uncertainty that are independent of the changes to market uncertainty. This means the metric used should not be a factor causing changes in market uncertainty. One good example described in the text that follows is the ability to forecast the market. When industry experts try to predict future markets, they are not changing the market uncertainty. Technological change is a poor metric because it is one of the factors that cause market uncertainty. When the architecture of Private Branch Exchanges (PBXs) used to provide voice services to business users changed in the 1970s to a program-controlled design that increased the vendor's ability to experiment with new features, this created market uncertainty. Suddenly customers had a tremendous choice among innovative features that were brand new, and because of the learning curve with new technology, the market uncertainty became high.

Even though it is difficult, estimating market uncertainty is important for showing a relationship to management structure. Fortunately, previous research by Tushman [3] explores how to measure market uncertainty in terms of forecasting error, which is the ability of industry analysts to predict the industry outcomes. McCormack [4][5] discusses the existence of a dominant design as evidence of low market uncertainty. This previous work shows the validity of estimating market uncertainty in the research community. Using a combination of existing and new techniques to estimate market uncertainty adds confidence to its measurements. What follows are the techniques used in the case studies from Part Two to estimate market uncertainty:

Ability to forecast the market.   Tushman's method explored a measure of how predictable a particular market is. He used historical predictions about markets and compared these to the actual market. The voice case study uses this technique by comparing PBX market forecasts to what the market was, showing low market uncertainty in the mid-80s. The ability to predict market trends and behavior implies low market uncertainty because it shows a basic understanding of the market. This metric is independent of market uncertainty.

Emergence of a dominant design.   McCormack discusses how the existence of a dominant design indicates low market uncertainty. The value of this method depends on the timing of when the dominant design emerged as compared to when the measure for market uncertainty is taken. As the dominant design is being determined, market uncertainty is decreasing as more users pick this design. Once the dominant design is established, its existence shows an independent measure illustrating a decrease in market uncertainty. The email case study in Chapter 8 illustrates this. The dominant design emerged in 1995 when major vendors such as Microsoft and Netscape implemented the Internet email set of standards. Once this happened, the uncertainty about what email standard to adopt was gone.

Agreement among industry experts.   Another indication of lower market uncertainty used in my thesis is agreement among experts about a technology and its direction. When market uncertainty was high, such as in the early email period, there was little agreement about the industry's direction and what features would turn out to be important to users. For example, how important were binary attachments and receipt of delivery to businesses for initial adoption? In the late 80s there was no agreement about answers to questions like this. In the early 1990s the experts converged on the idea of openness and interoperability, but they still did not agree on which standard would dominate. Finally, by the mid-90s there was agreement that Internet email was the future. Over time, agreement among experts is a measure for market uncertainty independent of changes in market uncertainty.

Feature convergence and commodity nature of a product.   In my thesis I discuss how when products are not differentiated by their features, they become like commodities, meaning that customers buy the product for reasons other than its features (such as price or the service policies of the vendor). This convergence of features demonstrates a metric similar to that of the dominant design. Initially, as features converge, there is linkage between the metric and the changing market uncertainty. After the convergence of features, and once the product has become more like a commodity, this metric is independent of changing market uncertainty. A fine example of this measure is the convergence of the features of PBXs in the mid-80s and how it transformed the market to a commodity nature.

Changes in standards activity.   My thesis examines how stable standards mean stable technology, thus implying low market uncertainty. Stable standards mean vendors have a fixed target on which to base products. Email is a good example of low market uncertainty after the standards stabilized. In 1996, the major Internet email standards (Mail, SMTP, POP/IMAP, and MIME) were established. Once standards for a particular technology become stable, we have a good indication of low market uncertainty that is independent of market uncertainty.

The preceding methodologies for measuring market uncertainty provide a reliable way to gauge market uncertainty at coarse granularity, indicating whether market uncertainty is low, medium, or high. These estimates provide a way to see significant shifts in market uncertainty. This is particularly true if several of the preceding methods agree.

There are many examples illustrating high market uncertainty. Even when a product is a sure market winner, the technology used is often unclear. Look at the VCR battle between two incompatible formats. The demand for home TV recording was clear, but the standard battle between VHS and Sony's Beta had much uncertainty. Many vendors and users picked the wrong technology and ended up with worthless investments in equipment when VHS emerged as the winner. Think about the videophone AT&T introduced at the World's Fair in 1964. Even with technology that makes this possible, users did not choose it, as many experts predicted would happen. A recent example of high market uncertainty are predictions about Asynchronous Transfer Mode (ATM) in the 1990s. A few years ago, it seemed that the answer to any question about networks was ATM; yet, although successful in particular niches such as multiplexing DSL links and WAN infrastructure, ATM never became a ubiquitous networking solution for LAN, WAN, and desktop network services, as many predicted. Instead, advances in Ethernet technologies enabled inexpensive, fast-switched Ethernet hardware that provided high-speed network access to the desktop at a low cost. These examples show that even when you know customers will embrace a particular product or service, picking the particular technology that customers finally adopt as a dominant design involves uncertainty.

The case studies of email and voice services presented in Chapters 8 and 9 use a combination of all the preceding methodologies to estimate market uncertainty using historic data. Voice services have seen several shifts in market uncertainty. At first, because of regulatory constraints, market uncertainty was very low, but this changed as regulation relaxed and technology advanced. With the maturing of technology came a commoditization of basic voice services. Now, technology is advancing faster with the convergence of Voice-over IP causing market uncertainty to increase. The history of email also provides strong examples of how to measure market uncertainty by looking for agreement with different methodologies.

Showing the level of market uncertainty using several methods gives a more confident estimate because each method uses different data. In these case studies, the sometimes-sparse data often came from secondary sources, making it important to find agreement for the different ways to measure the level of market uncertainty. The voice and email case studies are convincing because there is such agreement.

It is important to measure market uncertainty in current markets in order to make better decisions. In Chapter 10, I discuss how to do this by example with the Voice-over IP market, illustrating how industry pundits are in disagreement about what will happen. It looks at the great variety in features with IP PBXs, showing vendors' experimentation. Other current technologies are examined within this framework, such as wireless services in Chapter 11 and Web applications and services in Chapter 12. These examples show that it is harder to estimate market uncertainty in current environments than historical ones because there is less data to analyze.

Effect of Market Uncertainty on the Value of Experimentation

The economic value of experimentation links to market uncertainty by definition — market uncertainty is the inability of the experimenter to predict the value of the experiment. When market uncertainty is zero, the outcome of any experiment is known with perfect accuracy. As market uncertainty increases, the predictability of the success of any experiment's outcome is lower because outcomes are more widely distributed. This link between experimentation and market uncertainty is intuitive as long as the definition of market uncertainty is consistent with the variance of results from a set of experiments.

When market uncertainty is low or zero, the experimenter has a good idea of the market. This means that each experiment is expected to match the market well and meet the needs of most users. If market uncertainty is large, then the experimenter is unable to predict how the market will value the experiment. It may be a wild success (such as the Web) or a dismal failure, such as the attempt of PBX vendors to capture the business data LAN market in the 80s. Figure 6.1 shows how 100 experiments might be distributed on three examples of a normal distribution (each with a difference variance). These data points were simulated using an algorithm given in [6] for a normal distribution with mean = 0 and variance = 1, 5, and 10. This shows how the value of market uncertainty changes the benefit of experimentation; when market uncertainty is low (variance = 1), the best experiment has a value around 2.5 away from the mean. When market uncertainty is high (variance = 10), similar data yields a value of 25 away from the mean, an order of magnitude better than the low-variance case. This shows how, in an absolute sense, as market uncertainty increases, so does the possibility of performing an experiment that is a superior match to the market, as indicated by a value far above the mean of the distribution. It illustrates that when market uncertainty is low, even the best experiment is not far from the mean, but high market uncertainty disperses the outcomes over a greater distance from the mean.

The results of this simulation match what is expected for a normal distribution, as illustrated in Figure 6.2. It shows the probability of experiments being a particular distance from the mean. This matches the results from Figure 6.1 of a simulation with 100 experiments with different variances. Looking at the percentages in Figure 6.2, it is expected that 34 percent of these experiments will fall between the mean and +1 standard deviation from it, but only 2 percent of these experiments will range between +2 and +3 standard deviations from the average, which is confirmed by the simulated data in Figure 6.1. You expect to need more than 769 experiments to find a single service instance that has a value greater than +3 standard deviations from the mean.

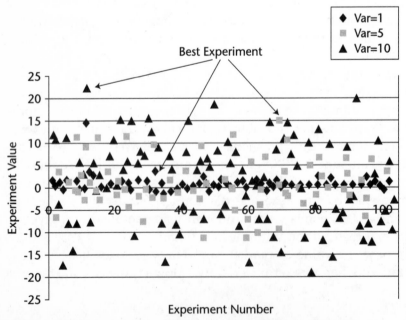

Figure 6.1   Link of market uncertainty to value of experimentation.

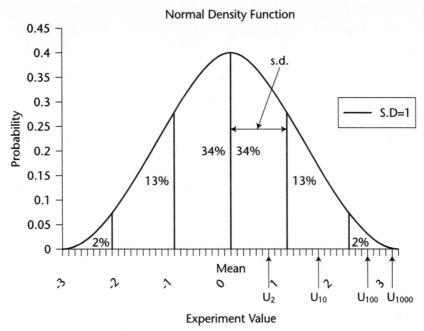

Figure 6.2   Normal distribution and experimentation.

In Figure 6.2, the expected values of the best experiment of 2, 10, 100, and 1,000 are indicated as: **U2, U10, U100,** and **U1000** (as calculated by Baldwin and Clark [7]). Again, this matches the results in Figure 6.1. As expected, roughly 2 percent of the results are between +2 and +3 standard deviations from the mean, and the best of these 100 experiments is around 2.5 standard deviations above the average. As expected, for each additional experiment the margin gain is less, which shows the diminishing advantage for each extra experiment. The intuition behind this is that, as you increase the number of experiments, the best of these experiments has a value that grows further from the mean, but at a decreasing rate. For example, with two experiments you expect one of the outcomes to be greater than the mean and within the first standard deviation of it. It takes roughly 10 experiments to expect the best result to be greater than 1 standard deviation from the mean, 50 experiments to be over 2 standard deviations from the average, and around 1,000 experiments to exceed the mean by 3 standard deviations.

The difference between the mean of the distribution and the best experimental result grows as the standard deviation increases. This implies that guessing right in markets with high market uncertainty produces greater value than guessing correctly where market uncertainty is low. In these examples, the distribution is normal, and the experiments are uncorrelated (the Random Walk model). My argument holds no matter what the distribution or correlation between experiments. As you get more experiments, or as the variance increases, then the difference between the average of the distribution and the expected value of the best of all the experiments increases. I model the Random Walk approach as the worst case because it allows an equal probability of either a very bad result or a very good one. When experimenters have more direction, you just expect better results.

There is an argument about whether experiments with new technologies are directionless, as indicated in Figures 6.1 and 6.2, or have a focus. This was a point of contention in my thesis defense, with Bradner and I defending the Random Walk theory against HT, who believed in focused experiments. This idea is explored in Kaufman [8], who also believes in the Random Walk theory. As pointed out previously, the answer to this does not hurt my argument. The likely answer is that experiments are not completely without focus, but the focus is less than one would like to believe.

Conclusion

Managers who don't understand uncertainty are at a disadvantage in the uncertain world of today. Investors who understand the uncertainty of user preferences in network-based services are better able to profit from this uncertainty by thinking with an options point of view, as explained in Chapter 5. High market uncertainty is the environment in which experimentation thrives and promises the most money to those that predict this market. The high uncertainty means that some winners can win big, which keeps things exciting. When experimentation has huge potential payouts it is likely that more players will be in the game; this is a good thing because having more players improves the odds of finding a superior market match.

The next chapter puts all the pieces together by explaining the theory of this book in detail: how market uncertainty increases the value of distributed management. It first states some assumptions about what group of network-based services this theory is useful for. These services must have business and technical reasons that imply central management is the right choice, but because of the difficulty of meeting uncertain markets, distributed management provides services with the most value. Next, the statement of the central theory in this book links the effectiveness of management structure to market uncertainty. It explains how high market uncertainty implies greater value of distributed management structure because it allows experimentation and how low market uncertainty means that central management is the best choice. The next chapter finishes up Part One and leads into Part Two with the detailed case studies of email and voice services.

Theories about How to Manage Network-Based Services

Managers and investors who understand at an intuitive level how market uncertainty links to the value of a particular management structure are better able to understand how users' needs are best met in uncertain conditions. The theory proposed in this book explains that distributed management is a better choice in uncertain markets because it enables more experimentation, even allowing users to create new services. It explains that when users' needs are well understood, the centralized structure is likely to work well. This theory represents a framework allowing a new type of analysis that focuses on the effects of market uncertainty on the choice of management structure for network-based services. Understanding this theory is a critical success factor for the modern manager and investors who must make decisions with incomplete information in uncertain conditions.

Previous chapters outlined the basic elements of this theory, explaining how market uncertainty is one important factor in the determination of how to best manage network-based services. In Chapter 2, the attributes of management structure are explained. Chapter 2 illustrated how centralized management has many business and technical advantages but tends to limit innovations because of the difficulty of experimentation — it may even be impossible for users to experiment. It explained how distributed management allows easy experimentation, even by users, but how it may not be very efficient in its use of resources, as well as being harder to manage. This concept of distributed management is expanded in Chapter 3, which discusses the ideas behind the end-2-end argument, and its relationship to distributed management structure. Then, in Chapter 4, the discussion about management architecture focuses on the effect of management structure on the ability to experiment and on who can experiment. Chapter 5 presents a discussion about options at an intuitive level, providing a baseline for thinking about this theory that links the value of management structure to market uncertainty. Chapter 6 discusses market uncertainty and how it changes the value of experimentation by showing that when uncertainty is low, all experiments are closely clustered around the mean, but that high uncertainty implies that the experiments have a wide range of values, some very good, some average, and some very bad. The pieces of the puzzle are now organized and ready to be put in place.

In the text that follows, the puzzle pieces are put into place by building a set of assumptions and theories based on the previous chapters of this book. First, a few assumptions explain the type of network-based services of interest and the nature of experimentation with these network-based services. Then, a few simple theories discuss the likely success of network-based services based on their management structure and the level of market uncertainty. These theories link the amount of market uncertainty to the type of management structure most likely to succeed in producing services meeting uncertain user needs. Next are assumptions about how market uncertainty might change, followed by a theory about how changes in market uncertainty cause a migration of management architecture better fitting this new level of market uncertainty. The chapter ends by completing the puzzle with the results of a real options-based model, allowing visualization of these assumptions and theory. This completed picture illustrates the theory of this book linking market uncertainty to choice in the management structure of network-based services.

This chapter presents an intuitive high-level view of a theory developed in my Ph.D. thesis. The appendix presents a more complete statement of these assumptions, theories, and a mathematical model based on the assumptions and theory representing the value of a network-based service, as a function of the amount of market uncertainty, the advantages of central management, the number of experiments the users have to pick from, and for how many generations the service is expected to evolve. This model is based on the theory of real options and illustrates the preceding trade-offs in a graphical form that helps managers and investors visualize the trade-offs. The model quantifies the intuitive arguments of this section with a simple set of equations based on previous work about real options.

This theory does not help managers decide how to manage all network-based services. Sometimes the choice of distributed or centralized management structure has nothing to do with the level of market uncertainty. The technical advantages for end-2-end encryption illustrated in Chapter 3 argue for a distributed management structure no matter what the uncertainty is. There are many services, such as email and voice, as illustrated in Chapter 4, where the choice of what management structure to use is complex and depends on the market uncertainty of the particular users. The services this theory helps understand are those services where there are advantages to centralized management structure; yet, when market uncertainty is high, distributed management seems to be adopted by an increasing number of users in these services.

This theory is intended to provide a framework to analyze the value of network-based services in the context of their management structure within the network. It is not intended to give absolute numbers or rules, but to show general relationships between market uncertainty, parallel experimentation with market selection, the benefit of centrally managed services, and service evolution. This theory is contingency-based and similar to an argument by Lawrence and Lorsch [1] showing that the best way to manage a business depends on the business. Similarly, the best management architecture for a network-based service depends on the particular users, at the particular time, because it should be based on the current conditions.

In the text that follows, I state a formal set of assumptions and a theory based on these assumptions that clarifies the value of network-based services architectures, such as end-2-end structures, that allow easy, simultaneous experimentation. This distributed end-2-end architecture is compared to architectures that provide a more efficient way to manage the service, but where experimentation is harder to accomplish because new services require changes within the network core. I believe that if conditions match those set out by these assumptions, then this theory is a reasonable representation of the trade-offs involved when deciding how to manage network-based services.

Theory

This theory assumes that the value a provider of a service receives is random because of market uncertainty. The theory illustrates how allowing users to pick the best approach from many ways to provide a similar service provides a good chance of achieving a service with features that are a superior market match. The theory accounts for technical and management advantages from the centralized architecture for network-based services, comparing them to the benefits of easy experimentation with the distributed management structure. The theory states that when the advantages of the more centralized service architecture outweigh the benefits of many experiments, a centralized management structure may be justified. Finally, the theory accounts for how services evolve from generation to generation. It states that at each generation of a service, service providers learn from the previous generation about what will work better for the next generation.

Here are two fundamental assumptions and a simple theory:

ASSUMPTION 1   **The market demand for network-based services has market uncertainty. This means that service providers (which includes entraprise users) are unable to accurately predict the value they will receive for providing a service. This market uncertainty is the subject of Chapter 6.**

ASSUMPTION 2   **Experimentation with services is possible, and a market exists to value the experiments. The value of a particular experiment is the success of its adoption. This experimentation is used to determine what service best matches the current market conditions in the context of what features will be the most popular.**

THEORY 1   **The expected value of the best of _n_ simultaneous attempts at providing a service is likely to exceed the expected value of any single experiment. As _n_ increases, the possibility of a truly outstanding market match grows. This result is illustrated in Figures 6.1 and 6.2 in Chapter 6.**

One way to view Theory 1 is in the context of options — having a choice is analogous to having an option. This theory demonstrates the value of network architecture promoting many experiments, compared to management structure, where experimentation is more difficult. One example of this is the choice between two standards for Voice-over IP, SIP, and megaco/H.248. As discussed in Chapter 10, SIP allows both end-2-end applications and applications with a more centralized structure, but megaco/H.248 does not. Theory 1 illustrates the value of protocols, such as SIP, that allow flexibility in the management structure they allow applications to have.

What follows is a set of stronger assumptions allowing a deeper theory considering the management structure of services based on the degree of market uncertainty. It defines more precisely how services have different management structures and how each structure has different values and attributes.

ASSUMPTION 3   The payout to the service provider offering the best of *n* choices is nonlinear. More experimentation and greater uncertainty increase the expected value. The service provider receives this value by providing the service that best matches the market. This is explained in Chapter 5 on options.

ASSUMPTION 4   The less disruptive and less expensive it is to develop and deploy a service, the more experiments there will be. Experiments in networks with infrastructure allowing applications with end-2-end architecture requiring no alteration to the network infrastructure are generally less expensive and less disruptive than environments where a more constraining centralized architecture requires infrastructure change and permission, as explained in Chapters 2, 3, and 4.

ASSUMPTION 5   For some services there exist business and technical advantages (BTA) that push providers to offer services that are more centrally managed.

Services with the property of Assumption 5 are interesting because high market uncertainty implies that distributed management will produce services with the most value, even given that it will be more expensive to provide these services because of the inefficiencies of distributed management. If the market uncertainty for these services is zero, then a centralized management structure will work best.

Assumption 5 focuses on services that have advantages to a centralized management structure. If centralized management has no advantage, then it is never a good choice for a particular service. The choice of best management structure for services interesting to us is not clear because of the trade-offs between the efficiency of centralized management and the benefit of more experimentation. With services like this, the situation is counterintuitive because services that are managed inefficiently do better in the market than services with a more efficient management structure. This book concentrates on situations such as this where the arguments supporting centralized management are strong, yet distributed management is a better choice.

Next is a discussion considering under what conditions the preceding advantage of experimentation and choice is not enough to outweigh the inefficiencies of managing a distributed service.

THEORY 2 If high market uncertainty causes the difference between the expected value of the best of *n* experiments and the expected value of each individual experiment to exceed the business and technical advantages of the centralized management structure, then a service provider should consider providing this service with a more distributed managed architecture. When market uncertainty is low enough that the advantage of having *n* choices is less than the business and technical advantages of a more centrally managed service, then providing the service with centralized management architecture makes the most sense.

Theory 2 demonstrates the value of a management structure that promotes experimentation by users compared to that of a management structure that allows experimentation only by centralized authorities. One way to view Theory 2 is to compare the value of many users experimenting to the value of one central authority undertaking a single experiment. It is market uncertainty that determines which has the most value: the efficiency of centralized management when market uncertainty is low or the flexibility of distributed management when market uncertainty is greater.

So far this theory looks at a single generation of a service, yet services evolve over time. Each generation has many attempts (experiments) to provide a service offering with a good market match. Thus, each service generation is composed of many service instances from simultaneous experimentation (that is, a group of services), which are the efforts of many different contributors. This theory incorporates service providers learning from the previous generation of experiments, thus reducing the market uncertainty from generation to generation.

ASSUMPTION 6 Those experimenting and providing services learn from experience, causing a decrease in market uncertainty.

THEORY 3 If market uncertainty is decreasing, then service providers are likely to succeed at providing a service with centralized management when the advantage of market uncertainty and parallel experimentation no longer outweighs the business and technical advantages (BTA) of the centralalized management architecture.

Market uncertainty can also increase if technology changes, as Clark [2] points out, leading to the following:

ASSUMPTION 7   Technology changes and alters the space of what services are possible, which causes market uncertainty to increase. One example of this occurred when PBXs became computerized — the whole scope of possible features changed, as discussed in Chapter 9.

THEORY 4 If market uncertainty is increasing, then service providers are likely to succeed at providing a service with distributed management when the advantage of market uncertainty and parallel experimentation outweighs BTA.

Theories 3 and 4 imply that market uncertainty is catalytic in the migration of users from one management structure to another. When market uncertainty is increasing, then the ability to experiment, which distributed management allows, becomes more valuable than any advantages to the centralized management structure. A reduction in market uncertainty, though, causes the shift of management structure in a network service from distributed to centralized. Reducing market uncertainty implies that the business and technical advantages of centralized management become more important than the benefit of increased experimentation from distributed management.

This theory is fundamental to understanding how to design the infrastructure used to build services based not only on business and technical advantages, but also on market uncertainty. It provides a framework to analyze the value of network infrastructure in the context of the type of management structure this infrastructure allows. It illustrates the trade-offs between centralized and distributed management structures with respect to market uncertainty, the number of experimental attempts to provide the service, how many generations the service evolves for, and the advantage of centrally managing the service. It shows that when a centrally managed service has an advantage from a business and/or technical perspective, the market for the service may still be better met with services that have less central management but allow more innovative features because of experimentation. It illustrates the value of flexibility in network infrastructure with respect to the type of management structure services can have.

Model Results

The preceding theory is hard to visualize without a mathematical model illustrating the tradeoffs described. The results from this section are from a mathematical model developed in detail in the appendix, which is based on the preceding theory. This model is similar in nature to the options-based approach by Baldwin and Clark [3] that explains the value of modularity in computer systems design. It expands on previous papers by Gaynor and Bradner [4][5] explaining the advantages of standards designed with modular structure because the standards increase the choices designers have.

This model focuses on two main forces affecting the value providers receive for services rendered: the benefit of many parallel experiments, combined with market uncertainty, pushing services to a more distributed management structure, and the efficiencies and other advantages of centralized management pulling services to centralized architectures. The model is based on the premise that environments providing easy experimentation may not offer the optimal management structure, and environments that are optimized for efficient service management may not be conducive to numerous experiments.

Figure 7.1 (see the appendix for the derivation of this figure) is a surface representing the additional value of a user having n choices for a particular service. It shows the value of experimentation along with its relationship to both market uncertainty and the number of experimental attempts for the service. This value is the marginal benefit from n experiments over a single experiment. As expected, the value to the user of being able to pick the best service from n choices increases at a decreasing rate with respect to n, the number of experiments. It increases at a constant rate with respect to MU, the market uncertainty. The surface illustrates this by showing the value of the best of n experiments (Z-axis), n (Y-axis) as the number of experiments, and MU, the market uncertainty (X-axis). The curved lines for increasing n show the decreasing rate of increase while the straight lines for increasing MU show the linear increase with regard to MU. This surface is a visualization of the value of users having a choice and how market uncertainty affects this value.

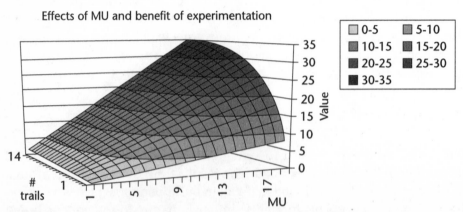

Figure 7.1 Market uncertainty's effect on the value of experimentation.

This result demonstrates that as market uncertainty increases, allowing users to provide services becomes more attractive because of the enhanced value of experimentation. If the cost differential between end-2-end (distributed) and centralized management structure is less than the benefit gained from high market uncertainty and parallel experimentation, then the value of the best application from all users and service providers is likely to exceed the value of a single attempt to provide the service by a central authority.

Figure 7.2 (see the appendix for its derivation) shows this relationship linking MU (the market uncertainty) (Z-axis), BTA of using a central management model (the business and technical advantage transformed into a cost differential) (X-axis), and n (Y-axis), the number of experiments run in parallel. This surface shows the relationship for a range of n (number of experiments) between 1 and 20. Points on the surface illustrate where market uncertainty is such that either architecture works well; points above the surface represent a region where the end-2-end architecture works better because of the advantage of allowing many users to perform parallel experiments, combined with market uncertainty. Points below the surface have low enough market uncertainty relative to BTA that the controlling central authority is able to meet market needs with a single attempt. The forward edge of the surface shows the amount of MU required to offset BTA for a single experiment. From here, the surface slopes sharply down with regard to the number of experiments, showing the great value of experimentation. As expected, the range of services benefiting from end-2-end type architectures grows with more experimentation. In addition, as expected, this growth occurs at a decreasing rate. The rate of decrease levels out quickly, at about 10 experiments, showing that the biggest marginal gain from parallel experimentation is from relatively few experiments. This surface represents a visualization illustrating how market uncertainty changes the value of distributed management.

The preceding results provide a framework that helps us understand the relationship between market uncertainty, many parallel experiments, and the advantages of a centrally managed service. It illustrates how high market uncertainty increases the value of the end-2-end management structure. The next result explains how services evolve from generation to generation and how learning from the previous generation reduces market uncertainty in the current generation. Learning occurs when service providers gain experience from the previous generation about the preferences in the market by watching experiments from other service providers. The next results create a more dynamic model, factoring in how services evolve as market uncertainty changes.

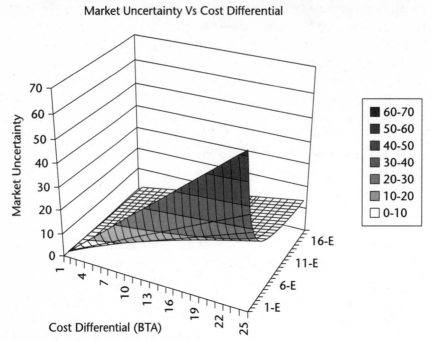

Figure 7.2 How to manage services.

Learning has the effect of lowering the benefit of many experiments because the standard deviation is reduced, causing each experiment to fall within an increasingly narrowing range centered around the mean; thus, many experiments help less and less. To model learning, from past generations it is assumed that at each generation the market uncertainty is reduced by a given amount. For example, suppose the market uncertainty is 1 at the first generation: If there is no learning, then the market uncertainty remains 1 at each generation. With learning it might be reduced by a factor of 2 at each generation. This means the market uncertainty is 1 in the first generation, one-half in the second generation, and one-fourth in the third generation. This model assumes that learning is symmetric — all service providers learn the same for all experiments run by everybody. The effect of learning changes the dynamics of how services evolve, causing migrations of users from one management structure to another.

Accounting for how a service evolves adds another dimension to the preceding figures. By holding the number of experiments fixed at 10, a similar surface to the one in Figure 7.2 is created, illustrating how the choice of management structure changes over time. It demonstrates the trade-offs between market uncertainty (Z-axis), advantages of centralized management (X-axis), and for how many generations (Y-axis) the service is expected to evolve, which enables a longer-term view of the management structure of the service. This analysis is depicted in Figures 7.3 and 7.4. Points on this graph have the same interpretation as those in Figure 7.2; points above this surface imply that distributed management is best, while points below imply that the advantages of centralized management outweigh the value of experimentation. Figure 7.3 is the base case, showing how a service evolves with no learning, while Figure 7.4 illustrates learning at the rate of dividing the market uncertainty by 2 at each generation. These figures show a very different picture of how services will evolve in the context of what management structure will work best and when to migrate to a new management architecture.

Figure 7.3 shows an example with no learning; it illustrates that centralized management is never the best choice if the service is allowed to evolve for many generations. Market uncertainty never decreases, so at each generation the gain from experimentation increases the total value of the best service. No matter how large the advantage of centralized management, if the service evolves enough generations it is overcome by the benefit of experimentation. It may take a while, but in the end, distributed management will win out.

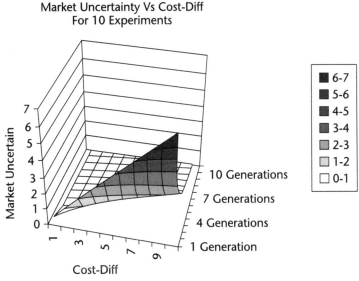

Figure 7.3 Evolving network-based service.

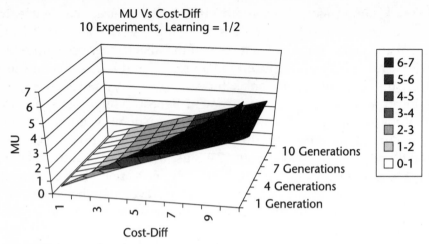

Figure 7.4   Evolving network-based service with learning.

Figure 7.4 represents a far different situation because the value of experimentation is limited. In this scenario, at each generation the market uncertainty is one-half that of the previous generation — at some point market uncertainty is so low that experimentation has little value, implying that the advantages of centralized management overpower the gain from experimentation. This surface converges to a limit illustrating that in this case experimentation is of value only for the beginning generations because of the rapid decrease in market uncertainty. Market uncertainty quickly becomes so low that even if the service is allowed to evolve forever, the gain from experimentation is small relative to the advantages of centralized management.

These graphs provide a framework to examine the trade-offs between market uncertainty, any advantages to a centrally managed service, and how many generations the service is expected to evolve, for a fixed number of experiments. These surfaces help answer the important question of which management structure works best when, by showing when the advantages of central management overcome the benefits of experimentations enabled by distributed management. (See the appendix for more detail.) Basically, this analysis helps to determine at which generation in the evolution of a service the advantage of many experiments becomes small compared to the management efficiencies of centralized network-based services, or vice versa, when the advantages of experimentation start to exceed the benefit of the centralized management structure. These surfaces help management understand when migration to a different management structure makes sense.

The results of this model point toward a two-tiered structure that network-based services live within. Figure 7.5 illustrates this structure — an outer region behaving like an incubator, growing new services, and the inner region, where the successful services from the outer region migrate when market uncertainty is low. On one hand, the ability to provide end-2-end services is necessary to meet user needs in uncertain markets. The ability to try out many different types of services, allowing the market to select the one with the best fit, provides superior service. After understanding consumer needs better, however, the ability to migrate the services into the network becomes necessary to capitalize on the business and technical advantages of centralized management. The outer region gives the power of innovation, while the inner region allows efficient implementations of the best ideas. Network infrastructure enabling both the end-2-end and the centralized management structures promotes this type of two-tiered structure.

As new technologies are created, and as users and service providers learn about these technologies, it is expected that users will migrate between centralized and distributed management structures. New technologies need the outer region of this figure to take advantage of experimentation. As the technology matures, the inner region becomes more attractive. The choice of management structure preferred by most users for services having long histories, such as voice, is expected to cycle back and forth between these regions. It is the combination of learning, which reduces market uncertainty, pulling services to the inner region, and new technology, which increases market uncertainty, pulling services to the outer region, that creates this dynamic cycling effect.

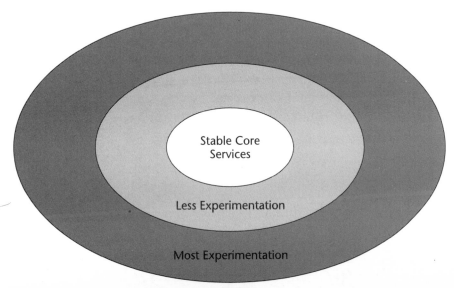

Figure 7.5   Two-tiered management structure.

Conclusion

This chapter ends Part One — the building blocks of managing network-based services. This chapter puts the puzzle together by knitting the previous chapters in Part One into a theory explaining which type of management structure is best for a particular network-based service within a particular market. This theory implies the high value of a management structure that allows everybody to experiment with new services when market uncertainty is high. It also illustrates how low market uncertainty reduces the benefit of experimentation, causing the advantages of centralized management to become attractive. This theory provides a framework for evaluating the choices management and investors must make.

The next chapter is the start of Part Two — the case studies of email and voice services, illustrating which management style was adopted and when. It looks back in time at email and voice markets and how management structure has changed over time. It correlates the degree of market uncertainty to the effectiveness of one management structure compared to another as measured by user preferences. These case studies look at the success of different management structures for the same service based on market uncertainty. They illustrate that the previously mentioned theories agree with the evidence in many important ways. They back up this book's theory and illustrate its power by explaining shifts in management structure unexplainable by other theories.

The case studies in Part Two illustrate how when market uncertainty is high, distributed management, such as PBXs for voice and ISP-based email, is adopted by many users. They show how low market uncertainty induces a shift from PBXs to Centrex, and from ISP to Web-based email. Voice and email, two very different services built on networks with very different infrastructures, have a similar evolution in the context of how market uncertainty has helped shape the evolution of their management structures. They provide solid evidence of this book's theory linking market uncertainty to choice in management structure.

The Case Studies

In Part Two, case studies illustrate the ideas discussed in Part One about how market uncertainty changes the value of experimentation and thus affects the choice between distributed and centralized management structure. Part One explained the framework linking market uncertainty to management architecture; in Part Two, the evidence is presented to convince you that this framework has value. The way that management structure evolved over time with voice and email services demonstrates the link between market uncertainty and whether users are adopting distributed or centralized management architecture—in both email and voice cases decreasing market uncertainty is the most likely cause for a shift to more centralized management structure. The cases that follow show how the theory in Part One explains the past with more accuracy than previous theories.

I validate the link between market uncertainty and management structure by correlating shifts in management structure to changes in market uncertainty. Concentrating on shifts in user preferences and then correlating these shifts to changes in the market uncertainty illustrates a possible link between market uncertainty and choice of management structure. In several examples the evidence supports a reduction in market uncertainty as the trigger in user migration from a distributed to a centralized management structure. This causality is demonstrated by ruling out the other plausible reasons for a shift in user preferences except for this reduction in market uncertainty. There is always a cost associated with changing management structure, which implies that users need a reason to migrate to a service with a different management structure than the current solution. The case studies in this section show that changes in market uncertainty can be the reason for this shift.

These case studies are not intended to be complete in nature, but instead focus on periods when users changed preferences in regard to the management structure they preferred. The email case study closely examines the emergence of large centralized web-based email service providers such as Hotmail and the effect this had on more-traditional email service providers with their more distributed management architecture. The voice case study traces how user preferences have shifted from distributed PBXs to centralized Centrex. The shifts in management architecture preferred by users in both email and voice markets fit the theory in Part One well.

Case Studies of email and voicemail

Email and voice services are two real-world examples illustrating the theory expounded in this book. Each case study begins with a general brief history of the service, followed by a more narrow analysis of one particular shift from distributed to centralized management structure. In both cases the evidence suggests that a reduction in market uncertainty at the correct time accounts for the shift in management style. In both cases there are many factors, such as technology change and regulation, which are possible causes for the shift in management structure. However, by careful analysis these other factors are ruled out as likely triggers for the shift. This leads to the conclusion that the decrease in market uncertainty was the most likely cause for this shift to more-centralized management structure.

The email case focuses on the shift of users to centralized email services such as Hotmail, which started in 1997. The case demonstrates that market uncertainty significantly decreased, as indicated by several different techniques at the time this shift occurred. It argues that other factors such as Internet technologies that change rapidly and new superior technology such as window-based email interfaces, the Web interface, and the Web itself are unlikely causes of this observed shift from ISP email to more centralized web-based email systems. The evidence indicates that reduced market uncertainty is the most likely factor to have caused the shift to centralized email services in the late 1990s.

The next case examines in detail the shift from the distributed model of PBXs to a more centralized Centrex service offered by the phone company. In the PBX market, regulation played an important role in shaping the market and the players within it. However, the shift in management structure from PBXs to Centrex in the mid-1980s is unlikely to have been triggered by regulation. Rather, regulation only exacerbated the degree of the shift. Centrex succeeded because as market uncertainty decreased the telephone companies could meet user needs by migrating already successful features from PBXs into the core of the PSTN. It became possible to meet user needs with centralized management structure once market uncertainty decreased enough so that the phone companies knew what features to provide.

These two cases have a common thread, showing a shift from a distributed to a centralized management structure triggered by a reduction in market uncertainty. This evidence is powerful, given the many differences between email and voice services, and differences in the infrastructure of the underlying networks on which these services are built. The migration of voice features occurred within the intelligent PSTN (Centrex vs. PBX). Email services built on the distributed end-2-end architecture of the Internet have seen a similar migration of users from distributed ISP-based email to centralized web-based services. The likely cause of users migrating in both these cases is a decrease in market uncertainty, which shows that the theory generalizes to many services, on networks with different infrastructure. The case studies illustrate the theory that a decrease in market uncertainty caused a shift to a more centralized management structure that utilized the resources better, had technical advantages, and because of low market uncertainty meets the market well. Two examples, each from a very different type of network, yet having a similar shift in management structure triggered by decreasing market uncertainty, suggest the generality of this argument.

The next two chapters in this part are the case studies. First is email, then voice. The research leading to these chapters is from my Ph.D. thesis.

CHAPTER 8

Email Case Study

This chapter presents a case study of the network-based service email, from its early days within the research community in the 70s through its stunning success in the late 90s, when the number of mailboxes grew to hundreds of millions in the United States alone. Initially, as many vendors competed for the dominant design, much experimentation occurred. Because market uncertainty was high at that time, this experimentation was of great value. When Internet email won and the standards stabilized, the popularity of more centralized email services such as Hotmail grew. Roughly 40 million more mailboxes currently exist for centralized email services than the more distributed (and traditional) ISP-based email systems. This case illustrates the relationship between market uncertainty and management structure — when market uncertainty was at its highest in the 80s, the value of experimentation caused the distributed management structure to work best. Later, as the maturity of the technology caused market uncertainty to decrease, the value of the centralized management structure overcame the advantage of end-2-end architecture. The emergence of a dominant design and stable standards indicates this reduction in market uncertainty. The history of email as presented in this case fits the theories of this book well.

The theory presented in Part One predicts the evolutionary pattern of email in the context of the standards that became accepted [1][2] and the way the implementation of these standards unfolded in light of market uncertainty. At first, market uncertainty was high; many competing email architectures existed with both centralized and distributed management structures. Each vendor offered a different feature set, allowing customers many choices among different email services. As this theory predicts, when market uncertainty was high, distributed architecture was more popular; as market uncertainty decreased, users migrated to a more centralized management structure. As this theory also predicts, the ultimate winner of the game (IETF Internet email) allows both distributed and centralized implementations of the standards, thus enabling it to prosper in any environment. Internet email is the architecture that allows the most experimentation due to its end-2-end nature, the openness of IETF specifications, and the modularity of those specifications.

History

There have been several different generations of email, as depicted in Table 8.1, with each generation influencing the services and architecture of the next generation. The rows of this table are the attributes of the email systems that existed in the particular generation. In the research generation, academics experimented with email service. Then, in the geek generation, email became popular with technical professionals exchanging messages. Next, in the business generation, the business community discovered that email could speed the flow of information and cut transaction costs. Finally, email became a way for the average person to communicate in the masses generation.

Table 8.1   High-Level Email History

GENERATION	RESEARCH (70S)	GEEKS (80S)	BUSINESS (90)	MASSES (95)
Systems	IETF, OSI	IETF, OSI, Post Office (many other proprietary systems)	AT&T, MCI, IETF, OSI, proprietary systems	IETF
Management	Distributed	Distributed, centralized	Centralized, distributed	Distributed and centralized
Market Uncertainty	Very high	High	Medium	Lower

In the early 1970s computers were expensive and networks few; however, email existed within the Internet for a select group of researchers. The 80s brought changes as computers became less expensive and more common in the workplace, with telecommunications and networking technology coming of age, creating both the need and the ability to build email systems. By the late 80s, vendors and service providers were experimenting with many different ways to provide email systems. Even the U.S. Post Office saw the potential and planned to offer email service, but the FCC did not allow it [3]. Both open (X.400, IETF) and proprietary (AT&T, MCI, IBM) solutions existed, giving users many choices. As the 90s arrived, it seemed (at least to the pundits) that sanity had come to the world; the ISO X.400, based on open standards, allowed users on heterogeneous networks and computer systems to communicate. It was an open standard, allowing vendors and service providers to implement it. It did have competition from the Internet, but few believed the Internet (also an open standard, but with a very different standardization process [2]) to be a credible threat because of the overwhelming acceptance of X.400. This acceptance by vendors, users, and governments, though, did not translate into products that customers wanted. By the middle of the 1990s it became clear that X.400 had lost to Internet email, which emerged as the dominant design. Internet email was victorious for many reasons, as discussed in [2], including the initial greater complexity of X.400. Compared to a very simple initial Internet email standard, the better standardization of the IETF compared to the ISO (at least in my opinion), and the success of Unix along with the availability of open source Unix-based email implementations. At about the same time, the MIME standard, which was created by the IETF in 1992 for encoding arbitrary (binary) content within email, came into widespread acceptance. The dream of interoperability and useful content exchange between users became reality: The dominate design was clear, and standards stabilized to indicate low market uncertainty. Web-based email started its explosive growth at this point. It permitted efficient centralized management and could interoperate with any other Internet email server — the best of both worlds. By the end of 1999, these centralized email services were meeting the needs of the greatest number of users, as shown by the fact that they had the most mailboxes [4]. The following timeline is a summary of this history.

1973   Start of Internet email with RFC561.

1981   Simple Mail Transport Protocol RFC788.

Office Protocol for mail server client in RFC918.

1984	Post Office Protocol (RFC918) for email client. Market totally unpredictable (size and structure). AT&T, ITT, GTE, MCI, RCA, WUI, Tymshare, GEISCO, IBM, IETS, ISO offer email. All systems are incompatible.
1990	Until now most mail intercompany; with arrival of X.400, long-term vision becomes clear.
1991	Move to X.400 a welcome sign.
	Most email vendors adopted X.400
1992	Widespread conformance to X.400.
1993	Internet hits business radar; services like AT&T, MCI, and OSI-based implementations have slow (or no) growth; Internet is growing fast.
1994	Gardner Group predicts SMTP should not be used for business because the lack of delivery receipts does not provide reliability.
1995	Win95 supports Internet email.
1996	Web browsers support SMTP/POP; Web-based email starts.
	Standards-based systems cheaper, uncertainty lower; MIME finished in November.
1997	SMTP/MIME is the only viable option for email.
1999	SMTP is the only choice; everybody is Internet-enabled.
	Few are running Sendmail on their desktop.

Internet Email History

The history of Internet email matches the pattern that the theory in this book predicts. Its beginnings were humble: RFC561 [5], written in 1973 (only a few pages), describes how to format email messages in terms of the body of the email and its meta information (specifically headers). In the early days, email had no transport protocol, but instead used FTP to transfer mail files. Only in 1981 did Simple Mail Transport Protocol (SMTP) become a standard with RFC788 [6]. In these early days Internet email was a true end-2-end application; devices inside the network did not know about the email application, as Figure 8.1(a) illustrates. Each host providing the email service directly sent and received email. This distributed structure made experimentation easy because any user could do it.

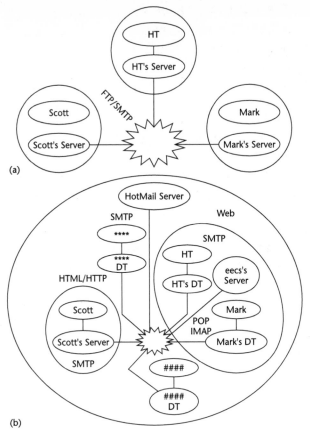

Figure 8.1 Evolving Internet email management structure.

Two main factors started driving a more centralized implementation of Internet email. First, users needed to receive mail on PCs that were not running all the time; second, not everybody had the hardware resources or technical experience to run email servers on a desktop. The Internet community responded to this demand with Post Office Protocol (POP) in 1984 with RFC918 [7] and then Interactive Mail Application Protocol (IMAP) with RFC1064 [8] in 1988, as illustrated in Figure 8.1(b). These protocols define a mail server that can serve many users on different hosts, without each individual host running a mail server. Instead, one host acts as a mail server, serving many clients running POP or IMAP on different hosts. The server remains up to receive and send email; clients need to access the server only when requesting email services, such as receiving or sending a message. IMAP, a later standard than POP, allows a more central management structure. Both POP and IMAP are more centralized in structure than the pure end-2-end solution because users of POP or IMAP are most likely affiliated with an ISP or company.

As PC usage increased, the limitations of ASCII-only email became apparent: Users had programs and data files not easily sent as email. This need caused the IETF to create the Multipurpose Internet Mail Extensions (MIME) in RFC1341 [9] in 1992, changing the very nature of how email was used. It was no longer just a way to exchange quick text messages — now all types of files were sent among users. This was the final piece needed for Internet email. With MIME, any type of information could be encoded, sent, and decoded into the original file, complete with meta information such as file type and name.

With MIME, all the main components were in place for a successful centralized implementation of Internet email, one that managed the clients' email messages for them and serviced a broad user base. The Web provided a nice platform for this, and in the mid-1990s, centralized Web-based email service providers such as Hotmail and RocketMail started providing service. These systems built with Internet email standards could instantly communicate with all other Internet email users; they were not isolated islands.

Analysis of Email History

The 30-year history of email has seen many different architectures and implementations. There are email systems based on a company's proprietary technology (for example, early versions of systems from AT&T and MCI) or on open standards that any vendor can implement (such as IETF Internet email or OSI X.400 email). Furthermore, systems built from the same standards may have a variety of management structures. For example, Internet email in the 70s and early 80s had an end-2-end structure. Hotmail, a service becoming popular in the 90s based on the same IETF standards, on the other hand, had a more centralized management structure. At first users had choices between what set of email standards to use; now the choice is between different implementations of the Internet set of email standards.

Based on privately owned technology, proprietary email systems tend to create a closed community where users can send email only to other users of the same system. These early systems based on proprietary technology created many separate islands of email users. Everybody on the same island (or in the same community) was able to send each other messages, but users in other communities (living on different islands) could not communicate with each other. In the 80s, before a dominant design emerged, many systems were of this type. Computer vendors such as IBM, large service telephone companies such as AT&T and MCI, Telex service providers such as Western Union International (WUI), and small private service providers such as Vitel (a company I worked at for 15 years) tried to become dominant players, but none of them did. This closed-community approach did not work well because nobody captured enough market to become the de facto standard; instead, it created many isolated islands of frustrated users who could not send email messages where they wanted to.

The theory presented in Part One predicts low innovation with proprietary systems, such as those just described, because only the owner of the technology can experiment with it. If market uncertainty is high, these private systems will have difficulty meeting user needs because of the limited experimentation implicit with proprietary systems. Users could not try out new ideas; doing so was counter to the centralized control imposed by some of these early systems and just impossible because vendors were not about to give out source code to their customers. One reason why these proprietary systems did not succeed during this early period of high market uncertainty is that users could not experiment with them.

Unlike proprietary systems, Internet email did allow users to experiment. In the early days of Internet email, many users made contributions to the standards and the base of free software to implement these email standards. Because Internet email is based on open standards and these standards started out simple, users found it easy to add new features and test them out. This is how standards for transporting email (SMTP) and sending attachments (MIME) were created; users made the important contributions. One reason Internet email became popular is that users could experiment with it.

Open standards allow any vendor or service provider to implement the technology. This technology may be free (as in the IETF) or have a cost (ISO X.400), but, in any case, it is not owned or controlled by any private party. One essential idea behind these systems is interoperability; that is, independent vendors must be able to implement email systems based on the open standards that work together, such as with IETF Internet email, or the ISO X.400 standards. Open standards do well when market uncertainty is high because anybody who is interested can experiment with them. The evidence that open standards meet users' needs well is the success of the Internet.

In addition to the standards defining email addressing, transportation, and content formatting, there are different architectures for implementation. Consider the two extreme ways to provide email based on the Internet standards: the end-2-end model, where each desktop runs a mail server, or a more centralized single mail server providing service for everybody, which can be used by anyone with a browser and Internet access. This example illustrates how to provide email with two very different architectures: end-2-end or a large core-based service, both derived from the same standard and interoperable. The end-2-end architecture allows more innovation; the centralized structure offers lower cost of management and other advantages, as discussed in Part One.

Management Structure for Email

As discussed previously in this chapter and in Part One, there are many different ways to manage email services, from distributed to centralized. In the most distributed architecture, email users run their own individual mail server. At the other extreme, every email user could access a single massive email server. Alternatively, email users could group themselves into units with a common tie (for example, working for the same organization, using the same ISP). Figures 8.2, 8.3, and 8.4 illustrate these two extremes as well as a middle-of-the-road architecture. As explained in Chapters 2, 3, and 4, the more distributed the management structure, the greater the innovation, and the more centralized the architecture, the greater the technical and business advantages.

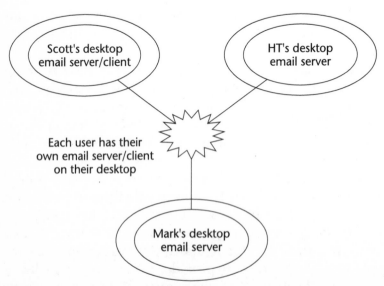

Figure 8.2 Distributed email structure.

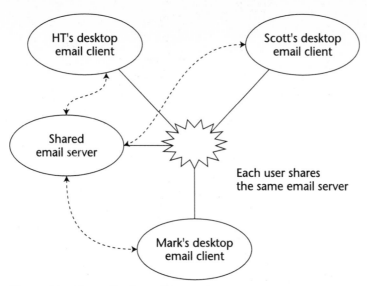

Figure 8.3   Centralized email structure.

Figure 8.2 shows the completely distributed architecture with every host acting as an email server. This is incredibly flexible for the expert user. It is easy to experiment with; it takes just two willing and able players to try out a new feature. The problem with this extremely distributed structure is its scalability. Imagine the complexity of managing email addresses with billions of mail servers. The cost to manage and maintain all these mail servers is just not feasible. Yet this is exactly the structure that Internet email began with, as discussed in this chapter.

The other extreme management structure for email is the single massive server illustrated in Figure 8.3. This is an efficient solution; you don't even need to transport messages between different email servers. Unfortunately, a single email server — along with the implied single email service provider — is not ideal because such environments tend to limit a user's options due to the lack of competition. Users end up losing out when they have no choice — the higher the market uncertainty, the greater the loss.

A more reasonable architecture is a middle ground, illustrated in Figure 8.4. Here users are grouped according to natural affiliations. For example, I work at Boston University (BU), and my email provider is BU. I also have a cable modem with MediaOne and the associated email account at MediaOne. Many users could just guess my email and be correct (mgaynor@bu.edu). This sort of structure is a very powerful incentive to users because it makes management so much less complex.

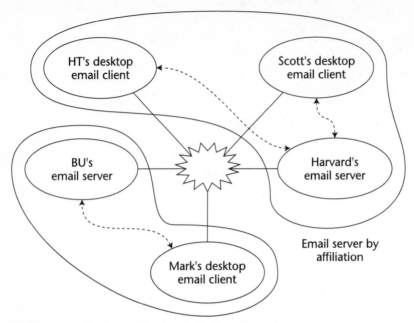

Figure 8.4   Mixed centralized and distributed email structure.

One important factor about management structure of email is who is responsible for email message management. Does the end mail user manage his or her messages on the local system that he or she reads mail on, or are these messages managed by the service provider? Figure 8.5(a) illustrates a centralized structure where email messages are stored and managed by the service provider. This is similar to how Hotmail works. This strategy is also an option with ISP email providers that support IMAP, as illustrated in Figure 8.5(b). There are advantages with this type of centralized management of email content — the user has access to all of his or her messages from anyplace in the world, and the user has less to worry about. Figure 8.5(b) also shows a more distributed model of email message management — let each user manage his or her own messages. This method has some advantages: It allows access to old messages without Internet access, the user controls the content (including backups of it), and this approach scales well on the server side. The nice part about Internet email standards is that they allow both centralized and distributed management of email messages.

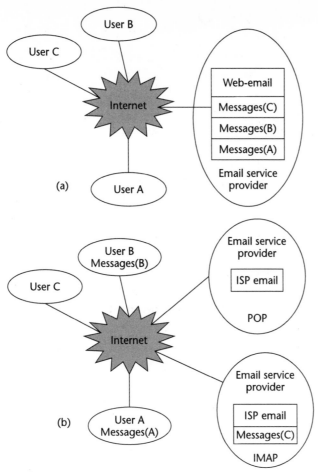

Figure 8.5   Message management.

As email history shows and as current usage illustrates, all three structures can and do interoperate. Some diehard users really do run email
servers on their desktop with themselves as the only email client reading
email sent to their private email server. Many users have email accounts
that identify them with either an ISP or an organization. The most popular
type of email service today, however, is the massive centralized architecture; think about how many people you know who have email addresses
ending in hotmail.com or yahoo.com.

ISP-Web versus Web-Based Email

ISP-based email systems, such as MediaOne Roadrunner, and Web-based email, such as Hotmail or Yahoo!, are based on the same Internet standard and have a similar Web-based user interface, but they are fundamentally different from a management point of view. The distributed management of ISP email (even when it uses a Web-based interface) contrasts with the centralized management of Web-based email. This is true from the scope of management viewpoint and the data management style. Both systems are completely compatible with Internet email specifications and are able to send and receive messages from and to each other and any other Internet email system. This split of email into centralized and distributed implementations gives users a choice, providing a good test for the theories in this book.

ISP-Web email is what most users have bundled with their ISP Internet account. There are many examples of this, including giant ISPs such as AOL and UUnet, cable providers such as MediaOne and USA, and dial-up ISP accounts bundled with a PC such as Dell-Net and MSN-Net. The address of the user's email reflects the ISP the user is associated with — for example, <user_name>@aol.net or <user_name>@mediaone.net. This provides a lock-in, making it expensive for the user to switch ISPs because the email address must also change. Most Web users are paying for ISP email because they need the ISP for basic Internet access and have no way to unbundle the service.

ISP-Web email has a distributed management style for two reasons. First, users of ISP-Web mail must be members of the ISP. For example, I have a cable modem with MediaOne; it is my ISP and provides me with email, basic IP, and a default Web portal page. Users may have several ISPs, with each ISP associated with an email address. This one-to-one association between your ISP and email account is distributed in nature because each ISP manages its email service independently of the others. ISPs can be very small or very large. Second, consider the management of data. Most end users of ISP-based email manage the messages on their own systems[1]. This makes it hard to use several computers for accessing email because you need to transfer email to different systems. Both the management scope and management of data have a distributed structure with ISP-Web email.

---

[1] AOL is one exception to this; it manages the users' messages. AOL is the most centralized of the ISP-type services because of this.

In contrast to this distributed nature of ISP email is the centralized management structure of Web-based email systems. These services, such as Hotmail and Yahoo!, provide email service to users of any ISP because the email account is not linked to any particular ISP. This means the user can change ISPs but keep the same email address. This implies a centralized scope of management. User messages of these Web-based email systems are managed on the email server, not on the users local system. (Note that IMAP does allow ISPs to offer users the option of letting the ISP manage the users' email messages, as Figure 8.5(b) illustrates, but this is not widely implemented.) This makes it easy for users to access their email from any system, anywhere, as long as they have Web access. Both the management scope and management of data have a centralized structure with Web-based email.

The comparison of ISP-Web mail to Web-based email is a good example to test this book's theory, which links market uncertainty to the value of a centralized versus a distributed management structure. Both systems provide a similar user interface via the Web. Most users of Web-based systems also have ISP-Web email bundled with their Internet service. These two types of email give users a choice of systems with similar service and user interface, but very different management structures. The growth of Web-based email has been explosive. By 1999, more Web-based email boxes existed in the United States and internationally than the total number of ISP-based email accounts [4], illustrating that users are choosing the more centralized management structure of Web-based email. This theory implies that when market uncertainty is low, the centralized structure of Web-based email will work well for more and more users.

Evidence of Shift

The evidence of email shifting to a more centralized management structure is the growth of Web-based email mailboxes and the number of different services offering Web-based email services. The growth of Web-based email shown in Figure 8.6 is phenomenal, starting at close to zero in 1996 and growing to more than 280 million Web-based email boxes in 2001 [10]. This is greater than the total number of ISP email accounts. Web-based email now represents the greatest percentage of mailboxes, both domestically and internationally [4]. According to industry sources [10], there are so many free sources of Web-based email that compiling a comprehensive list is impossible. Based on what users are doing, the shift to centrally managed Web-based email is clear.

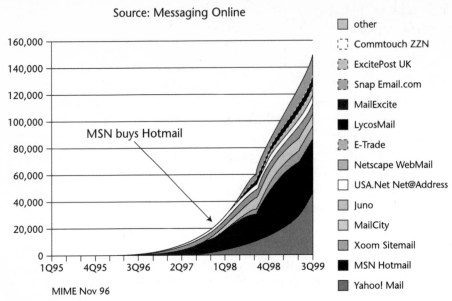

Source: Messaging Online

Figure 8.6 Web email growth.

A comparison of ISP email with Web-based[2] email shows the success of email with a centralized management structure. Figure 8.7(a) shows the growth curve for Web-based compared to ISP email from the mid 1990s to the fourth quarter of 2000. At first, Web-based email did not exist, but it grew quickly after its introduction in 1995 and 1996 by service providers such as USA, Hotmail, and Juno. This migration of users to a more centralized structure started in late 1997, before the rapid growth in ISP email caused by more Internet users. This shows that the shift to Web-based services started before the explosive growth of the Internet and the Web. By the end of 2000, there were more centralized Web-based email boxes than all of the ISP-based mailboxes combined. Figure 8.7(b) highlights this shift by showing the growth rate in mailboxes per quarter. It illustrates the rapid growth of Web-based compared to ISP-based email.

The biggest players have the lion's share of the market. For Web-based email services, MSN Hotmail and Yahoo! mail are the leaders, with roughly 50 percent of the market share [10]. For ISPs, AOL claims about 10 percent of the market share. Figure 8.8(a) shows the growth of AOL compared to that of Hotmail and Yahoo!. Note the difference in the growth rate between AOL and Hotmail or Yahoo!. AOL had flat growth, but both Hotmail and Yahoo! have much steeper curves, which illustrate the rapid shift to the largest centralized email services. The steep growth of services such

---

[2] The number of unused Web-based email accounts is hard to estimate, but the shift is still clear.

as Hotmail and Yahoo! compared to the shallow growth of AOL shows the magnitude of this shift in management structure. Figure 8.8(b) highlights this difference by showing the growth rate in mailboxes per quarter. The growth of AOL is constant, but the two biggest Web-based services show periods of explosive growth.

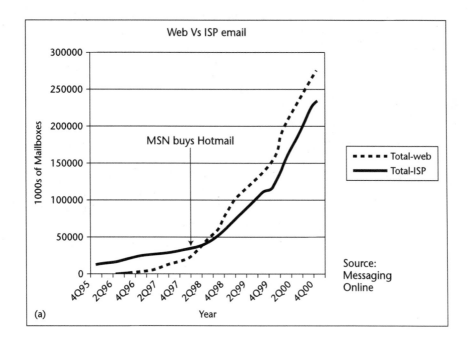

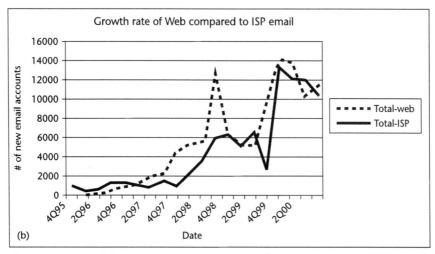

Figure 8.7 Email shift in management structure in 1997.

Why Users Migrated to Centralized Email

The shift of management structure in email is clear, but the cause is not. There are many possible causes for shifts in management structure besides changes in market uncertainty. Did shrinking market uncertainty cause the shift to a centralized management structure, or was something else the trigger? Could the technology of the Web and the Internet, GUI email software, Web browsing software, regulation, and the success of portal sites such as Yahoo! be catalytic factors of this shift in management structure? First, the decrease in market uncertainty at the correct time is explained; next, other factors are ruled out as the catalyst of this shift in management structure. This section demonstrates that a reduction in market uncertainty is the most likely cause behind the observed shift to a more centralized management structure.

Decrease of Uncertainty in Email

As expected with new technologies in the mid-80s, the email market was highly uncertain. John McQuillan, a well known pundit in the telecommunications industry and contributor to *Business Communications Review*, says, "It is extremely difficult, perhaps downright misleading, to offer forecasts of the structure of the marketplace, much less its size, beyond the next two to three years" [11]. Vendors and service providers included AT&T, ITT, GTE, RCA, Western Union, Tymshare, GEISCO, and IBM [11]. Each of these proprietary systems behaved like a walled garden; users of each system could send messages to other users of the same system, but not to users on different systems. Also in use was the Internet email suite, which started in 1973 with RFC561 [5], and the OSI/ITU X.400 standard, which was becoming an international email standard. In 1985, even industry pundits could not, and would not, predict the future, a good indication of high market uncertainty.

In the early 90s things became clear, at least to the pundits such as McQuillan, who writes "until now most mail was inter-company, with the arrival of X.400, the long term vision is clear" [12]. By 1991, the move to X.400 was welcomed by vendors who decided to adopt the standard [12][13]. Users, however, did not choose X.400, instead adopting the Internet set of protocols. This caused vendors to rethink their strategy. The pundits were half-right: They correctly predicted that open standards were the answer; they just picked the wrong open standard, a good demonstration that market uncertainty was still high.

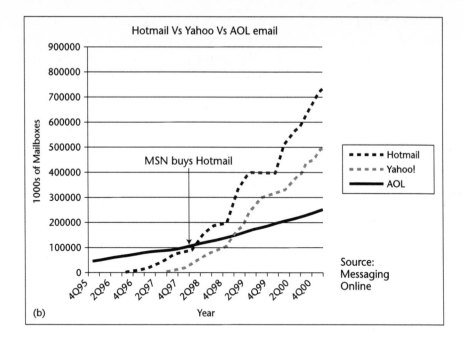

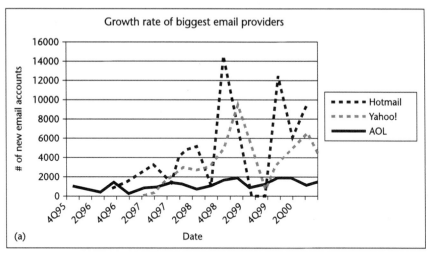

Figure 8.8 Growth of largest email providers.

In about 1993, Internet email hit the business radar because services provided by communication giants such as AT&T and MCI had slow or no growth, while Internet email use was growing quickly [14]. A 1996 Forrester report [15] discusses the importance of the major online service providers adopting the Internet mail protocols and how this opened the floodgates for interservice messaging. Forrester correctly believed that the Internet standards would replace proprietary standards within the next three years [16]. David Passmore wrote about the end of proprietary email and the success of Internet email in 1996 [17]. By 1999, Internet email was becoming ubiquitous [18][19] as it became the normal way many people communicated with each other. By the shift to a more centralized management structure in 1997, the established standard for email was the Internet suite. This acceptance of the Internet standards as the dominant design illustrates that market uncertainty was low.

Market uncertainty can be a state of mind: It is similar to the concept of customer confidence, a leading economic indicator used by the U.S. Commerce Department. When the market believes a thing will happen, many times it does. When both Microsoft and Netscape became Internet-email-compliant, it indicated a major step in the acceptance of Internet email. Now, virtually every desktop in the United States has software able to access Internet email. Microsoft was the important player because Internet email was part of Windows 95 — "the fact that Navigator included E-mail capabilities was a serious who cares for Windows 95 users, who now get E-mail software along with all of the other little doodads in their operating system" [20]. Other vendors, such as IBM/Lotus, Qualcomm, Apple, and Eudora [17], are supporting Internet email standards, showing further industry convergence to Internet email as the dominant design. This vendor confidence in Internet email convinced even the most conservative IT manager that Internet email was the dominant design and the only viable choice for email standards.

After the final group of MIME RFCs 2045–2049 in November 1996 [21][22][23][24][25], the framework for Internet email stabilized, indicating lower market uncertainty. This technique to measure market uncertainty requires studying how Internet email standards have evolved over the last 25 years. With high market uncertainty, one would expect many changes in the basic specifications, but with low market uncertainty, standards should be more stable because most changes tend to be small, incremental improvements. Figure 8.9 shows the history of how Internet email standards have changed, detailing how the standards for Internet email have stabilized over the years. It shows the main Internet mail RFCs and the increase of pages in each specification over time. It demonstrates how the format for mail (RFC561) [5] was settled by the early 80s. Next, it shows how SMTP (RFC788) [6], the email application transport protocol, became a standard in the early 80s. MIME started in 1984 with RFC1341 [9], becoming settled by 1996 with RFCs 2045–2049. Note the pattern: At first, the changes are larger and then shrink in size as the standard matures. Figure 8.9(a) shows the number of pages changed per day, and Figure 8.9(b) illustrates the rate at which these specifications are changing, as measured by the number of pages changed per day. As this graph shows, the final change in three major email specifications (MAIL, SMTP, MIME) all had a similar rate of change, which is 0.02 pages per day. This shows the reduction in market uncertainty based on the rate of change of email specifications, indicating that Internet email standards were becoming stable, thus implying lower market uncertainty.

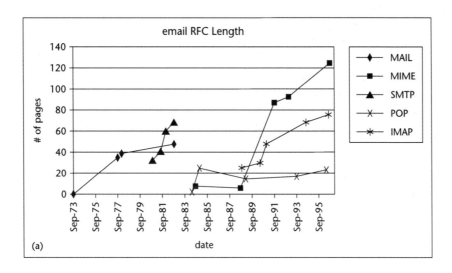

(a)

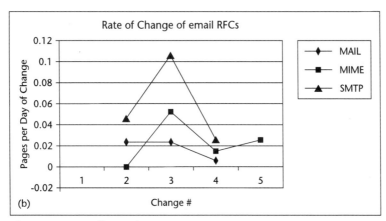

(b)

Figure 8.9   Email RFC changes.

Analysis of Other Factors

From the preceding text it is clear that market uncertainty changed at the correct time, but what other factors might be responsible for this shift? These factors include Internet technologies and the Web, window-based email user interfaces, Web browsers, and bundled ISP email accounts, such as MediaOne. All these technologies had some impact on Web-based email; however, the evidence suggests that these other causes were not catalytic in the migration of users to a more centralized management structure.

The technologies of the Internet, including email protocols and the Web, are certainly a necessary condition for Web-based email. Nothing was changing in these technologies in 1997, though, that could account for the observed shift in management structure. The Internet and the Web, while undergoing rapid evolution, had a stable base of standards providing the framework for interoperability. As discussed in the previous section, the major Internet email standards had stabilized by 1997. To the application writers, Web-based email was like many other database applications, consisting of a front-end Web-based user interface and a back-end database. It did not require the invention of new technology. The Internet technologies enabling Web-based email systems existed before the observed shift of the management structure in 1997.

Browsers and the prominence of Web traffic on the Internet occurred too early to cause the shift of management structure observed in email. Several years before the shift, browsers became popular and Web usage became a significant factor. By September of 1993, Web traffic became the tenth-largest source of Internet traffic [20]. This is less than one year after Mosaic released its browser [20]. Netscape's Navigator debuted in October of 1994. By April 1995, Web traffic became the number-one source of Internet backbone traffic. The popular use of browsers to access information on the Internet existed several years before the shift in email management structure.

Could the Web user interface be a contributing factor in the email management shift because of its consistency and friendliness? This seems unlikely because the Web interface is not unique to the Web — it's just another window-based application, like applications in Microsoft Office. Browsers displaying graphics are X applications in the Unix environment and Windows applications in the Microsoft world. The consistent user interface is a property of the windowing system, not of the application using it. Microsoft's Windows 95 became the standard GUI for the office desktop. Notice that the user interfaces on Microsoft's Internet Explorer, Netscape's Navigator, and the Eudora email system are similar to the look and feel of Microsoft Office. They all conform to the Window's UI and use the standard pull-down menus, scroll bars, and other expected items. Users had experience with the Web, Windows-based user interfaces, and GUIs for email several years before email's observed shift in management structure.

Unlike the preceding factors, other factors, such as the success of portal sites like Yahoo! and MSN, did occur during the management shift. Portal sites gave people a place to go on the Web; did they also provide the access to Web-based email users had been waiting for? It seems not, given the timing, and different types of successful portal sites, some with and some without email.

One might believe the success of portal sites such as Yahoo! and Hotmail gave email a boost; however, the opposite happened. Email helped portal sites such as Hotmail differentiate themselves from other Web search sites. Yahoo! did not announce plans to buy RocketMail until October 1997 [26], and by January 1, 1998, Microsoft decided to buy Hotmail and its then current 12-million mailbox base [27]. This occurred after the shift to centralized management that started in the beginning of 1997, as shown in Figure 8.9. There was no integration of Web-based email and portal sites when the shift started in early 1997; therefore, email boosted the value of portal sites, not the reverse.

The success of portal sites does not depend on email, any more than the success of email depends on portal sites. Informational sites, such as Cnet or CNN, that provide news are successful, yet they do not offer email services. Local portal sites, such as boston.com, are successful, yet do not provide email. Many Web indexing and search sites, such as AltaVista, Hotbot, Google, and Lycos, have succeeded in attracting users without offering email services. It seems unlikely that the success of portal sites caused this shift in email.

The preceding discussion shows how the two most successful Web-based email providers acquired email services only after their proven success. This follows this book's theory; after these services proved successful, big organizations started providing the service to cash in. Companies running portal sites bought successful Web-based services and integrated them into their portals as a way to strengthen user loyalty. This indicates that Web-based email services prospered before portal sites provided the service.

The changing technology of computers and the Internet, such as email GUIs, the Web, and the success of portal sites cannot explain the observed shift in structure. The reduction in market uncertainty, however, can. It was the correct time, and nothing else seems likely to have triggered the migration of users to centrally managed email. This illustrates that the reduction in market uncertainty is the most likely trigger for the shift in management structure seen in 1997.

Conclusion

Internet email succeeded as the theory predicts; it is modular in design, started out simple and evolved, allowing easy experimentation. Also as predicted, less flexible but more efficiently managed implementations of the dominant design (Internet email) became popular as the market settled. As expected, when market uncertainty decreased, innovation became less valuable than efficient management. Figure 8.7 illustrates this by showing the growth of centralized implementations of Internet email, such as Hotmail or Yahoo!, compared to traditional ISP email. This explosive growth illustrates how well a service such as Hotmail or Yahoo! meets the needs of the average user when market uncertainty is low. In the late 90s, users had a choice between ISP and Web-based email; many chose Web-based email because of the advantages of its centralized management structure.

The environment of the mid-1990s was complex due to the rapid pace of technology and the creation of entirely new types of businesses, such as portals and Internet search sites. Neither the introduction of new technology nor the success of any type of Web site caused the shift in management structure seen with email. The technologies for the Web, browsers, and email all existed before the shift in management structure. At the time of the shift, users could choose either system with equal ease, but they picked the centralized version more than 50 percent of the time.

The timing of this shift from a distributed to a centralized management structure fits this theory well. When MIME, the final piece for Internet email, became a stable standard, a centralized management structure could easily meet this fixed target. Furthermore, vendors believed this standard was the dominant design, as its adoption by Netscape and Microsoft shows. Because nothing else is likely to account for the shift in management style, and because market uncertainty had just decreased noticeably, it follows that this reduction in market uncertainty most likely triggered this shift.

The next chapter provides more evidence supporting the theories in this book; it presents a case study of voice services, with a focus on the shift from PBXs to Centrex in the mid-1980s. It illustrates how the theories in Part One explain the evolutionary history of the voice service market. It demonstrates that when market uncertainty was low, Centrex, with its centralized management structure, became more popular, and in conditions of high market uncertainty, users migrated to the distributed architecture of PBXs. Other theories have failed to predict the unexpected growth of Centrex in the mid 1980s; this book's theory explains that Centrex was a good choice at this time because of the low market uncertainty in the mid-1980s.

Basic Voice Services

This chapter examines the market for voice services over its long history. By understanding how voice services evolved, were adopted, and were managed, we might better understand how current similar network services will grow. The contrast and similarities between voice and email services provide powerful evidence to support the theories in this book — voice services are mostly built on the intelligent Public Switched Telephone Network (PSTN), while email services are mostly built on the dumb Internet structure. Voice services, similar to email, have seen shifts in the style of management structure that works best at any particular time relative to the level of market uncertainty. In both cases, high market uncertainty favors a distributed management structure, while low market uncertainty implies that a central architecture works best. The differences between the "smart" PSTN and the "stupid" Internet and the similar evolution of services within each network provide strong evidence proving the link between market uncertainty and choice of management structure.

Everybody is familiar with voice services over the PSTN because we use them daily in both our business and personal lives, which makes voice services attractive as a case study. The PSTN is the telephone network that switches most voice traffic. As discussed in Chapter 3, the PSTN is managed by large carriers and provides the voice services over a smart network not allowing end-2-end services, but instead promoting services that are centrally managed. While prohibiting true end-2-end services, the PSTN has evolved an architecture that allows distributed management structure for voice services — the Private Branch Exchange (PBX). PBXs are just telephone switches, similar to what is used at the Central Office (CO) of the PSTN, but typically smaller. They provide the ability to manage voice services locally. These PBXs evolved a rich set of features such as call transfers, caller identification, speed dialing, and voice mail. Because a PBX is similar to the switch in the CO of the PSTN, there is no technical reason why this switch located at the CO could not offer the same features as a PBX, so the telephone companies did this and called it Centrex. Centrex is a centralized managed version of PBX-like features that users found valuable. The comparison of Centrex to PBXs is similar to comparing home users who must choose between an answering machine located at their house (distributed management) or renting this service from the telephone company (centralized managed). In both cases, the user has a choice between a distributed and a centralized management structure for a similar set of services. Centrex and PBXs are two ways to provide similar services but with a different management structure, making voice services an excellent case study to test the theory from Part One.

The evidence derived from the history of voice services is compelling because it is the first ubiquitous telecommunication service with a long, rich history. This history fits the theories from this book well. When Centrex began in the 1950s, market uncertainty did not exist because users had only one choice: Bell dictated what products and services were available to users. Later, market uncertainty increased when regulations relaxed and technology advanced. By the late 70s most industry experts believed Centrex was dead because of the perceived advantages of the distributed management of the PBX — and for a while they seemed correct. Market uncertainty was high, and distributed management of voice services exploded — PBXs became a big and important market. The experts, though, were incorrect in the long run because instead of dying quietly, Centrex staged an unpredicted recovery when market uncertainty decreased as PBX technology matured. The decrease in market uncertainty made it easy for Centrex to offer PBX-like services that met the needs of all but the most demanding users, and to offer business and technical advantages as well. The revived and more competitive Centrex services became a threat to the dominant architecture — the PBX. The centralized management of Centrex proved to be what many users wanted, and it again became popular, despite recent reports in the trade press about the certain demise of Centrex. Finally, in today's market, the convergence of voice and data is again creating market uncertainty as IP-based PBXs begin to capture the attention of IT managers; organizations, such as Cisco, that create the technology for voice/data convergence currently deploy IP-based PBXs in most of their offices, proving the viability of this technology. Voice's long history has many shifts in management structure, and these shifts are positively correlated to changes in market uncertainty — when market uncertainty is high, the distributed nature of the PBX fits the needs of more and more users; however, when market uncertainty decreases, users shift to the more centralized management architecture that Centrex offers.

This chapter presents a case study of basic and advanced voice services as provided over the PSTN. Discussed first is the history of voice services, including the evolution of PBXs and Centrex service. The history for this case study starts with the introduction of Centrex by Bell in the 1950s. Next, in the early 1980s, the distributed model of voice services became popular as PBX features exploded. Shortly thereafter, the distributed PBX solution found unexpected competition from centrally managed Centrex. This shift to Centrex service started in 1983 and continued in force into the late 90s. This case study illustrates that market uncertainty decreased before users started shifting to the more centralized management structure of Centrex. Furthermore, this study concludes that other causes, such as regulation or technology change, are unlikely factors triggering this shift. By ruling out factors other than market uncertainty as the probable cause of this shift, this chapter concludes that, because market uncertainty was shifting at the correct time, it is most likely responsible for this observed shift in management structure. The case study of voice services matches the email case study, showing how a reduction in market uncertainty is catalytic to a shift in management structure from distributed to centralized.

For much of the early history of voice services, customers had little choice because, in the name of universal service, AT&T provided voice services in a monopolistic manner. FCC Commissioner Nicolas Johnson delivered the *Carterfone Decision* in 1968 [1], which resolved Dockets 16942 and 17073 in FCC 68-661 and stated that it was illegal to connect a non-AT&T device to the phone network. What started with *Carterfone* ultimately led to the Modification of the Final Judgment (MFJ) in 1984 [2], breaking up Bell and opening the market for equipment that provided voice and data services and over the PSTN.

PBX versus Centrex

Most business users have two main options available for voice service: the PBX located at the customer site and Centrex, a service provided by the phone company via a phone switch (for example, Nortel DMS-100 or AT&T 5-ESS) at the Central Office (CO). These different strategies are shown in Figure 9.1, where businesses A and B are served by PBXs on site, and business users C and D receive a similar service by sharing a part of the CO's central switch. Today, both services have competitive prices and meet most users' needs. For advanced users, PBXs have more innovative applications and give users greater control; for example, customers can schedule upgrades for new features according to their needs. With Centrex service, the user has less control and fewer features, but he or she does not need to spend resources or time thinking about maintenance, upgrading, and management of on-site equipment. Some users demand the local control of PBXs, but others find the business and technical advantages of Centrex more compelling than local control.

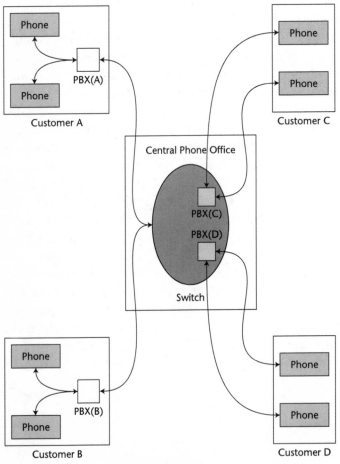

Figure 9.1 Centrex compared to PBX.

Comparing PBXs to Centrex service is similar to comparing Internet end-2-end applications to services such as Hotmail with a more centralized management structure. With PBXs, non-owners of the telephone network can experiment with new features, similar to the way in which Internet users can experiment with end-2-end Internet applications. A PBX vendor can implement a new feature for local use without changing the core telephone network. To add a new Centrex feature, the switch vendor must first implement the feature, and then managers of the telephone network must agree to upgrade their switch to provide the service. This is similar to comparing Hotmail to self-managed email — with Hotmail, only the centralized authority can change the features Hotmail offers, but with self-managed email, each organization can make changes as it wishes. PBXs offer a more end-2-end architecture than Centrex, similar to the way self-managed email is more end-2-end than Hotmail. As with the comparison between self-managed email and Web-based email in Chapters 4 and 8, PBXs permit more innovation and Centrex offers greater efficiency.

Created to provide the same service as a PBX within the CO of the telephone company, Centrex defines a centralization of very stable PBX technology. Centrex was successful until the PBX industry shifted from a step-by-step [3] mechanical design to the Stored Program Controlled (SPC) architecture [4]. This paradigm shift caused an explosion in PBX features because this new technology made it so easy to add features. Users in search of more innovative features flocked to PBXs. Soon PBX features converged to a standard set common to most vendors [5]. Then, as vendors and users understood these features better, they migrated to Centrex service, and users started to prefer this centralized solution. As advanced applications such as Automatic Call Distribution (ACD) and voice mail (VM) became popular both as standalone systems and as features integrated into PBXs, these applications moved into the network and became advanced Centrex applications. Centrex is a good example of how successful services from a distributed environment (PBXs) migrate into a more centralized architecture.

There are many reasons why innovation is greater with the distributed model of management that PBXs allow. These reasons include a greater number of vendors, the ease of experimentation, and the fact that there is less complexity and less likelihood that features will conflict. Innovation with PBXs first took place with basic voice features such as call forwarding and caller ID. Later the innovation moved to advanced voice features such as ACD, VM, and integration with computer databases via Computer

Telephone Interfaces (CTI). Innovation in PBXs has not stopped; even today, many businesses believe the most advanced voice applications are available on PBXs or as standalone attachments (for instance, Octel Voice Mail). The distributed nature of the PBX promotes innovation because it is easy for PBX vendors to experiment with new features.

Advanced Voice Features

PBX vendors classify voice mail (VM) and Automatic Call Distribution (ACD) as advanced applications because of the complex processing required to offer these services. While ACD and VM are very different from a technical point of view, they are considered two of the most important advanced applications for voice services in the last 15 years. Both VM and ACD applications helped keep the PBX market alive by allowing vendors to profit, something they were having difficulty achieving by selling basic voice features. By the 90s, most innovation occurred within these advanced features. These features first existed in standalone units, then became integrated into PBXs, and finally migrated to a Centrex service, but only after a market existed for them.

The function of an ACD is simple: routing incoming calls to operators. It is the ideal service for the telephone network to provide and fits well with the design of the PSTN. Given the intelligent nature of the PSTN, the network is the logical place to provide this service. Providing it at the network edge seems intuitively more difficult and less efficient because calls require rerouting after the network has already routed them. VM features are different[1].

There seems to be no technical advantage to providing this service inside the network. The information the network needs to provide this service is not related to its basic job: routing a call by setting up a circuit. To provide VM service, the network must save state information after a call is over. Even given the technical differences, ACD and VM features have evolved in a similar way. Each allows experimentation at the edges, and each sees a migration of successful features to Centrex and even residential service.

ACDs

Automatic Call Distribution (ACD) features are important to many different types of organizations, from large traditional companies, such as airlines, to smaller companies, such as niche ISPs or small mail-order houses serving customers by phone. Large users, such as major airlines, demonstrated the

---

[1] But, some features are better implemented with a network-based solution — for example, voice mail when the user is busy with another call.

need for this service and spurred the development of large, complex, and expensive systems. Then, as these systems became more affordable because of new technology, smaller niche users discovered that ACD features could save money and provide better service to their customers. A variety of organizations use ACDs and have different requirements because of the diverse groups of customers that they serve.

The function of an ACD is to route calls intelligently to a pool of operators. ACDs started out as standalone systems at the network's edge, but these features were soon integrated into the PBX. Finally, these features migrated to the network, thereby becoming a Centrex service. Many users are happy with the generic, centralized, network-based version offered by Centrex. This is not true for all users, though. Some advanced users need the more innovative features found through experimentation and the additional control offered by end-based units located on the customer's premises. When market uncertainty was high, there was no ACD Centrex, but as the uncertainty decreased, ACD features migrated inside the network, meeting users' needs because of low market uncertainty. As time passed, more advanced ACD features continued their migration to Centrex services, but the most innovative features still appeared in end-based systems.

Voice Mail

Voice mail (VM) is almost ubiquitous; most businesses use it, and most residential users have limited voice mail service. The definition of VM in this book is not precise. For businesses, VM is defined as one of the following: a standalone VM system from a vendor such as Octel, a PBX/VM integrated system by vendors such as Nortel or AT&T, or VM provided as a feature of Centrex service. These advanced VM systems provide more than *answering machine* features. For residential users, VM is defined as a more basic service, such as that provided by home answering machines or network-provided basic VM service. The focus in this chapter is on business use, not the large residential market.

In the beginning of the 80s, AT&T implemented several early VM systems, including a system for residential users and the ECS for businesses [6], but it was thwarted by regulation when the FCC declared VM as an *enhanced service* under the Computer II decision [7][8]. Owners of businesses providing manual, non-automated answering services wanted protection from AT&T and got it, or so they believed. Of course, answering services powered by human operators had a limited future because technology provided a better way to perform this function for many users. By the mid-80s, standalone vendors filled the niche AT&T was prohibited from serving because of regulatory constraints. At first, growth was slow. Users did not understand the features this new technology could provide and how these features could save time and money, thus providing a strategic competitive advantage. As vendors experimented with new features, and as customer expectations evolved, VM became a fast-growing option in PBXs. By 1988, regulations had relaxed [9][10], allowing Centrex to offer VM services, which it did successfully because of the low market uncertainty.

VM services first became successful with end-based architectures; then, as experimentation showed what worked best, PBXs integrated these features into their systems. Next, the most successful VM features migrated into the Centrex feature set when regulations allowed. As expected, Customer Premise Equipment (CPE) VM systems had the most feature-rich systems [10]. PBXs have nearly the same feature sets as standalone units; last, VM service within the network via Centrex meets the needs of many average users, but lacks the cutting-edge features found in more end-based solutions. Users' adoption of VMs matches the theory presented in Part One well — distributed architecture is more popular when market uncertainty is high, but when market uncertainty is low, many users migrate to centralized structure.

History of PBXs

In the mid 1970s, everything changed in the voice world; PBX architecture shifted from a purely mechanical, step-by-step design to a programmable architecture called Stored Program Control (SPC) [11]. This new paradigm enabled PBX vendors to experiment with new features as never before, and they did. Along with PBXs came smart phones with features better able to utilize the advanced features offered in PBXs. This shift in technology caused market uncertainty to increase because users and vendors did not understand the possibilities of the new architecture.

By 1980, the digital PBX[2] came into being. Pundits believed that this technology would merge voice and data, but it never happened. Instead, it turned out that the best way to provide voice services with the SPC architecture is digital technology. Later, this digital design enabled advanced applications such as VM and computer database integration. The digitization of voice was an incremental advance [12], and an intuitive one at that, because computers handle digital data better than analog signals.

---

[2] A digital PBX converts the analog voice signal into a digital pattern. This came after the SPC architecture, but it was a natural evolution of using computers to control PBXs because computers interact with digital data better than analog signals.

In the early 1980s, because it was so easy to create features with the new SPC architecture of the PBX, the number of features offered by vendors exploded; by 1984 most vendors had more than 500 features [13]. Fancy PBX features blew Centrex service out of the water [14] because the advanced abilities of feature-rich PBXs provided a compelling reason to switch. Smart phones took advantage of new PBX features such as displays showing caller ID or time on line, message-waiting lights, and buttons for speed dialing and auto-redial [15]. Some experiments were successful; others were not. For example, the PBX attempt in the early 80s to become the office LAN failed badly because it did not meet user needs. This period of experimentation helped vendors understand users better, causing a decrease in market uncertainty.

By 1986, things settled down in the PBX market. At the low end of the market, the PBX was a commodity [5]; at the high end, advanced features sold systems [5]. The majority of users found their needs met by lower-end PBXs, causing price to become an important factor in vendor selection, indicating lower market uncertainty.

By the early 90s, PBX vendors were troubled; the market was in the doldrums, and for many users the PBX had become a commodity. Low-end KEY (KEY systems are like low-end PBXs) systems and smaller PBXs had advanced features [16][17] and thus met the needs of the average user well. These features included fixed and programmable keys, LCD displays, transfer, conference, hold, last-number redial, speed dialing, message waiting, and even basic ACD and VM functions found in the most advanced PBXs of the early 80s [16]. PBX applications such as VM and ACDs were growing at more than 20 percent per year [18]. Most PBXs worked for most users, as shown by the price-driven market [19]; there has been little innovation over the last few years [20][21]. Systems were very similar [20][22], with features differing little from those developed in the 80s. Most users depend on six or fewer features [23].

In the 90s, the PBX market settled down; basic voice services were stable, and advanced features such as VM and ACD continued as the major growth area for vendors. The high-end market consisted of fancy VM, ACD, and call accounting features [24]. Some features became popular, such as ACD and VM (as predicted); other features, such as voice/data integration, failed to be adopted by users (not as predicted). Following is a timeline history of the PBX with particular attention to market uncertainty (represented in the timeline by roman text) and the migration of features (represented in the timeline by *italics*) from PBXs to Centrex.

1982	Flashy PBXs blow Centrex away.
	Phones with display giving caller ID, speed dialing, and more for new SPC PBXs become popular.
	PBX market has high uncertainty.
1984	Features proliferated because of SPC to more than 500 with most vendors.
	Expanded feature set develops in Centrex.
1986	Features in PBXs from most vendors are similar.
1987	*Centrex is becoming more competitive to PBX because of more PBX features.*
1988	PBX predictions of last few years wrong because they did not account for Centrex's unexpected resurgence.
1989	*Centrex ACD lags behind PBX ACDs.*
	PBXs *Digital Centrex introduced.*
1991	No new PBX technology introduced.
	PBXs becoming a commodity, market is driven by price — market is in doldrums, days of selling boxes are over, now vendors sell applications.
1992	PBXs have matured; systems are very similar.
	PBXs are not in telecom trade shows.
	Centrex is giving PBX vendors a run for their money because of competitive pricing and features.
	At high end, Centrex cannot compete with advanced PBX features.
	Only Nortel DMS-100 has ACD features, but most PBXs do.
1993	Feature-wise, PBXs are not much different from 80s, and most PBXs have the same features.
	Centrex is becoming much more competitive with features like ACD/VM.
1994	*ACD features such as queuing and alt-routing migrate into Centrex.*
1996	*Centrex is weak in regard to advanced call processing (CTI).*
	Centrex changes slow in reaching market because of the difficulty with changing CO switch software.

1997	*What's a telco to do: sell PBX or Centrex? They don't know!*
	PBX offers minimum of problems; the features are basically the same.
1998	PBXs are not the only option as Centrex meets needs of a growing number of users.
	Selsius-Phone has H.323 (Voice-over IP)-based PBX, not the vast features of PBXs but the basics like hold, transfer, and conference.
1999	Lucent, Siemens, and Cisco announce LAN-based IP/PBX.
	Cisco plans to use its new system in some offices. IP/PBXs will not have more voice features, but will be less expensive, "good enough," more flexible, and based on open standards so they will interoperate with equipment from other vendors.
	No real changes over the last 25 years to PBXs, but now? IP/PBX?

ACD History

By the early 90s, the integration of ACD features into PBXs increased. These features became more popular as businesses discovered the cost-effectiveness of ACD features and how to use these features to gain a competitive advantage. PBX vendors had money and incentive to provide these advanced services; this caused the feature gap to close between integrated and standalone units [25]. Low-end PBXs and KEY systems began to have basic ACD features [26], as ACD technology became more common and customers became more educated. By the mid 90s, PBXs were capturing more than 60 percent of the ACD market, but, for advanced ACD applications, the standalone vendors had 50 percent of the market [27], illustrating that some advanced users needed the cutting-edge features offered only in standalone units. The success of PBX vendors with ACD features forced standalone vendors such as Rockwell to be aggressive in order to keep ahead of PBX vendors [28].

By 1996, ACD features included links to Ethernet with TCP/IP stacks and Computer Telephone Interfaces (CTI) that linked database information with incoming callers [29] in a standard way. The most advanced features like CTI at first were available only from standalone vendors (Rockwell, Aspect), with PBX vendors such as AT&T and Nortel working to make them available [29]. The market continued into the 90s with ACD features shipped on more than 25 percent of low-end PBXs and more than 50 percent of high-end units [22], showing that ACDs became one of the most important features for PBX. The following timeline illustrates a history of ACD features in the PBX. Uncertainty is represented by the roman text and *italics* represent migration.

1970s	ACD is used for only the largest users, such as major airlines and the largest hotel chains.
1987	Market is $100 million in 1986, predicted to be $250 million by 1991.
1988	By 90s all PBXs are predicted to have some type of ACDs.
	AT&T PBXs have ACD features.
	Nortel announces ACD software for DMS-100 for Centrex ACD service.
	New PBX ACDs are beginning to match Stand Alones (SA) in features.
1989	ACDs are becoming more popular.
	Centrex ACDs lag behind PBX and SA units.
1991	*All major PBXs have ACDs.*
	Gap is closing between ACDs integrated into PBXs and SA, with greater differences between vendors.
	KEY systems getting ACDs illustrates the commodity nature of these features.
1992	Standalone units face intense competition from PBXs.
	Modern ACDs use CTI, but fewer than 800 are installed.
	Only tariffed Centrex ACD is on Nortel's DMS-100.
1994	*Features such as queuing and alt-routing are now in Centrex ACDs.*
1995	*PBX vendors have two-thirds of market, SA has one-quarter, but for advanced ACD applications, SAs have 50 percent.*
	ACDs are one of the most important features in last 10 years (along with VM).
	Rockwell/Sprint provides in-network ACD.
1996	ACDs link to Ethernet/TCP for CTI.
	Most carriers are offering network-based ACDs; they fall short of on-site units but are catching up with load balancing, time-of-day routing, and so on.

In 1996, 610,000 ACDs call agent positions shipped; 75 percent were new systems.

Voice Mail History

A study showed that in the early 1970s, 75 percent of all calls were not reaching the called party [30], highlighting a major inefficiency in the business world and providing a business opportunity. Some businesses used answering services that employed operators to take messages manually for unanswered calls. AT&T realized that technology provided a better answer and tried to sell both the at-home and ECS business system, but was halted by regulators responding to lobbying from answering service business owners who feared a collapse in their business model. Answering service owners, however, could not stop VM products from other vendors with regulation, so they lost anyway. By 1983, users had many options: traditional services, standalone units from vendors such as VMX and Octel, and a few PBX vendors that started to integrate this feature into their products [31]. At this point, growth with the automated VM systems was much slower than expected, with a market of $20 to $30 million [31].

In the early 80s, market uncertainty was high; most systems had fundamentally different architectures [32]. An open system, such as Ericsson's or Rolm's, allowed any incoming caller to leave a message for a mailbox owner. Closed systems, such as Bell's Dimension, allowed mailbox owners to leave only other mailbox owners a message, a design that proved of limited value. Much experimentation took place as vendors tried to figure out what features would appeal to users. Market growth was slow, as Figure 9.2 shows; by 1985, it was only $200 million [33]. Things were changing, though. By the mid 1980s, VM system cost was down from the $50,000–$500,000 range, to $15,000–$75,000 [33], creating a bigger market.

By the late 80s, market growth was good, and the regulatory environment for the Regional Bell Operating Companies (RBOCs) improved because they could now provide store-and-forward services, including VM [34]. The RBOCs are the pieces of AT&T resulting from Bell's breakup in 1984 caused by the MFJ. Uncertainty was decreasing, but it was still not low because this was only the third generation of VM systems [35]. The most innovation in fancy features, such as auto attendant and directories, still existed only in standalone systems [36][37]. Now users had even more choices: standalone units, PBX/VM systems, and Centrex.

By the end of the 80s, VM became the fastest growing feature set in PBXs, with roughly 40–50 percent of high-end PBXs having VM options [38]. Centrex was also successful, with 15 percent of these lines having VM service [10]. As expected, Centrex VM had fewer features and less control than standalone or integrated PBX units [10]. Market uncertainty was moderate because vendors and users still had a lot to learn.

As the 90s started, VM was growing, but still did not meet the predictions from the mid 1980s [39]. VM features moved to smaller PBXs and KEY systems, showing the commodity nature of the basic feature set. VM (along with ACD) were the applications reviving the PBX industry.

In 2000, as expected, users still had a high interest in VM, and vendors such as Nortel in its CallPilot product were introducing Web-based interfaces to their VM systems. VM is successful as a Centrex service to business users and in-network answering machines provide simple VM services to residential users. Standalone VM systems from vendors such as Octel[3] still offer the most advanced features, and Centrex still offers the most basic, with PBXs in the middle. The following timeline shows the highlights of VM evolution. Uncertainty is represented by the roman text, and basic facts are <u>underlined</u>.

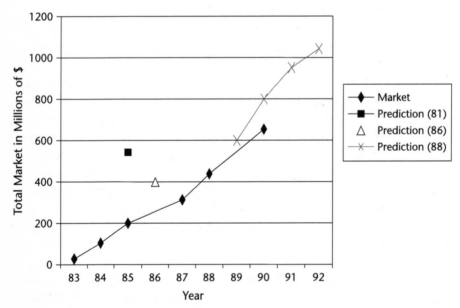

Figure 9.2 Growth of voice mail.

---

[3] Notes from a conversation with Nancy Kinchla, Director of Telecommunications at Harvard.

1981	Two VM systems — AT&T for home use and ECS for business — are introduced but halted by regulation.
	Study illustrates the need for voice messaging because three-quarters of all calls don't reach the called party.
1983	VM growth is much less than expected (market is now $20–30 million).
	There are many options for users.
	More PBX vendors are integrating VM, different systems (open compared to closed).
1984	Market exceeds 100M.
1985	Market exceeds 200M.
1986	Cost is way down for VM systems.
1987	Market reaches $290 million (less than predicted).
1988	RBOCs are allowed to provide VM.
	Less uncertainty occurs in the third generation.
	Rolm is the most aggressive PBX vendor.
	No standards are in place; all systems are different.
	Market reaches $433 million.
1989	VM is the fastest growing PBX option.
	Centrex VM (15 percent of lines) offers fewer features, less control than VM in PBXs.
	Of high-end PBXs, 40–50 percent have a VM option.
	Centrex is growing faster (6 percent versus 3 percent in PBXs), which implies more opportunity in Centrex.
1990	Market reaches $650 million (less than predicted).
1991	VM moves to KEY and small PBXs, showing the commodity nature of VM.
	VM predictions from the 80s prove to have been way too high.
1992	Fifty percent of Merlin Legend PBXs ship with VM.
1994	VM becomes one of the successful PBX applications.
1997	Interest in VM is still high.

1998	Web-based interface to VM becomes available.
2000	Octel SA VM is still competitive with cutting-edge features.
	<u>Applications like VM are generating profits for PBX vendors.</u>

New-Generation PBX

In the mid 90s, technologists talked about the death of proprietary PBX architecture in favor of an open client/server architecture that would allow lower cost and more flexibility. Would PBXs morph into voice servers [40]? Industry expert Richard Kuehn [41] believed traditional PBXs would live long, but some vendors bet that markets existed for open-architecture PBXs. For now, the main handicaps with open systems are the ultra-high reliability of traditional PBXs along with their highly evolved advanced applications and features.

In the late 90s, the market became more uncertain. Vendors liked closed systems because they lock in users, but users preferred open systems because they are more flexible and cost less. Voice-over IP was heating up, and because it is based on open standards, it pushed PBXs toward a more open architecture. In 1996, Mitel introduced a LAN-based PBX [42], and other vendors, such as Lucent, Siemens, and Cisco, began to enter the LAN-based PBX market [43]. This fortified users' beliefs that, in the near future, classic proprietary PBXs would not be their only option [44] as PBXs based on open standards became viable. For now, IP/PBXs may lack the hundreds of PBX features, but they do have the most desirable ones, as shown later in Table 9.2. Many believe this new generation of PBXs will be less expensive and more flexible and will work *well enough* for many users within a few years [45].

As the PBX market moves into the twenty-first century, traditional PBXs are more a commodity than ever. Only the most advanced users require the cutting-edge applications offered by new PBXs. The growth of IP-PBXs is slow for now, but it will increase as the technology matures and meets a greater number of users' needs. This new technology has revived interest in PBXs because of the inherent advantages of integrated voice/data networks.

History of Centrex

The creation of Centrex illustrates the centralization of PBX features when market uncertainty about their customer acceptance was low. In the 1950s, NY Telephone realized that it could save money by providing PBX features inside the CO [46] because the expense of installing, removing, and then reinstalling a PBX — on average every three to five years in a densely populated business district — was increasing. There was no technical reason not to provide PBX-like services within the CO, and there was economic justification for doing so. At this time, PBX technology was stable because the Step-by-Step design was mature, implying that the feature set to implement was a fixed target [11]. There was also no competition; AT&T provided all service and equipment. At NY Telephone's insistence, AT&T built Centrex service into a CO switch that NY Telephone installed in 1958. As my theory predicts, when market uncertainty was low, migrating PBX functions into the CO switch had cost advantages. Centrex worked well until deregulation allowed other vendors to compete in the market, and the technology of the SPC PBX allowed for easy experimentation with features.

Centrex services failed to keep pace with innovation in PBXs because the new SPC architecture of the late 70s [47] allowed such easy implementation of new features. For the largest users (businesses with more than 10,000 users), Centrex was the only solution [48] because PBX technology was unable to scale up to this many users. For smaller users, the feature set of new PBXs proved appealing. PBXs were hot and gave users many choices because vendors offered innovative features along with the fancy phones to utilize them. Analysts believed Centrex was on its deathbed because it could not compete with the PBX feature set [7][14][34].

The prophets of technology were wrong because Centrex was hardly dying. The mid-80s brought unexpected changes as Centrex expanded its features to be more competitive with PBXs [14]. By 1983, Centrex's rebirth was underway, and the number of new lines shipped per year reversed its downward trend [14]. It was an unpredicted revival as industry experts had written Centrex off. Centrex kept improving its feature set by implementing the services innovated in PBXs, such as speed dialing [7]. By 1985, the total number of Centrex lines installed began to increase. As the market rebounded, Centrex kept on implementing ideas that originated in PBXs, such as Digital Centrex [49], which helped Centrex grow throughout the 80s. By 1989, growth in Centrex was 6 percent, as compared to 3 percent in the PBX market [10]; this growth occurred mostly because Centrex was efficient, was easy to manage, and met the needs of many users. Centrex was growing faster than the PBX market, illustrating just how wrong the industry pundits were.

In the late 80s, Centrex starting offering advanced features seen only in PBXs a few years earlier, such as voice mail [10], the fastest growing PBX option, and ACD features. These advanced Centrex applications were not as feature-rich as in standalone units or systems integrated into PBXs. Centrex could not compete with high-end PBXs, but only a few of the most demanding users needed the sophisticated features found in the high-end PBXs.

Centrex got hotter [47][50] in the 90s, meeting users' needs better, as many popular PBX features from ACD and voice mail migrated into the network — features such as queuing and alternate routing [51]. Centrex still lags behind, though, in advanced applications such as Computer Telephone Interfaces (CTI) because Centrex developers and service providers don't provide new features as quickly as PBX vendors do. Centrex is still showing strong growth, but PBXs continue to have the edge in most advanced applications, with more innovation and a better ability to meet advanced users' quickly evolving needs. Centex still meets the needs of many, even though its advanced applications are not as innovative as new high-end PBXs. The following timeline shows this Centrex chronology. Uncertainty is represented by the roman text, *italics* represent migration, and basic facts are underlined.

1982	Flashy PBXs blow Centrex away.
1983	The downward trend of the number of new Centrex lines shipped reverses.
1984	*Expanded features for Centrex become available.*
1985	*Centrex is still more expensive.*
	Centrex is implementing more PBX features.
	Centrex rebirth gets stronger with the number of Centrex lines increasing — this trend is unpredicted.
1987	*Centrex becomes more competitive with more features and lower cost.*
1988	Digital Centrex and Distributed Centrex arrive.
1989	Growth in Centrex reaches 6 percent versus 3 percent for PBXs.
	Voice mail is the fastest-growing Centrex feature.
	Centrex ACD lags behind PBX ACDs.
	Services previously available in digital PBXs are now available with Centrex.
	Price is becoming competitive.
1991	Centrex users, overall, are satisfied.
	Lags exist between features available on the CO switch compared to what is installed in the current CO switch.
	Centrex gets 15–20 percent of smaller systems sales.

1992	<u>Centrex resurgence is still unpredicted and not under-stood by many; it caught vendors and industry experts by surprise.</u>
	Pricing for Centrex is competitive with pricing for PBXs.
	Centrex cannot compete with advanced PBX features, but most users don't need these.
	Only Nortel DMS-100 has ACD features.
1994	*More ACD features, such as queuing and alt-routing, migrate into Centrex.*
1996	*Centrex changes are slow to market because service providers don't upgrade to new software quickly.*

Over the last 20 years, Centrex became more competitive with PBXs in terms of cost and feature set. This cost evaluation is complex, though, and quantifiable only in a fuzzy sense given the nature of the costs involved. What is the value of having equipment on site? Sometimes, to the technically advanced and to companies with a business model that depends on cutting-edge advanced voice services, control is very valuable. To other businesses that need less cutting-edge voice applications, this value of control may be negative; the space to house the equipment might be more valuable than the benefit of having control over it. Another hard-to-quantify attribute is the reliability of Centrex over PBXs, which may depend on geography. Some users in California find that Centrex has more protection against earthquakes (and now power outages) than the organization could afford for its own equipment [52]. A comparison of Centrex to PBX pricing in the early 90s shows the competitive nature of Centrex and the complexity of the comparison [38][53]. Centrex is competitive to PBX ownership or leasing, but the analysis is complex and subjective.

Changing Management Structure

There have been several shifts in the management structure preferred by users with voice services. At first, users had little choice in how they managed their voice services because only AT&T provided these services and equipment. When Centrex emerged in the mid 1950s, the economics of the local phone company dictated who had what service. Everything changed in 1968 with the *Carterfone Decision* — now that they could, other vendors besides AT&T started building equipment. Users could now pick between Centrex and PBXs from several vendors. In the late 70s, technology altered the landscape of the PBX industry, as PBXs became programmable. This was believed to be the deathblow to Centrex; users had become enamored with the fast-growing feature set of PBXs. It was impossible for Centrex to keep up with the innovative new features that programmable PBXs allowed. Industry experts believed Centrex would never become competitive with computer-controlled PBXs; however, this proved wrong as users did start to migrate back to the centralized architecture of Centrex in the mid 1980s. This shift, along with some important historical events from the preceding timeline, is illustrated in Figure 9.3. These events illustrate several shifts from distributed to centralized and from centralized to distributed management structure.

At first, voice services shifted from a distributed to a centralized model. The first telephone call fits the definition of end-2-end applications spelled out in Chapter 3: two smart devices (phones) connected by a dumb network (the wire). The end-2-end architecture of the first telephone system does not scale because you can't connect every telephone to every other telephone — that would require too many wires. Telephone services fast became centrally managed; telephones were connected to a central switch, and the switch figured out what connections to make. This fits my definition of centralized management in Chapters 2 and 4 well — a telephone that was smarter than the wire in the end-2-end structure becomes the dumb end point in a more intelligent network (the switchboard). In the early days, this switchboard consisted of one person, the signaling being the caller telling the switchboard operator to whom they wanted to talk. The same telephone that seemed smart compared to a wire looks dumb compared to a smart switchboard operator; this highlights the point that the architecture of voice services shifted from distributed to centralized.

In the early 1900s, business users shifted to a more distributed management structure with voice services. The growth of businesses, coupled with the communication habits of business users within a large organization, created the need for the PBX. People within a single organization had extensive contact with others in the same company. The economics of central management did not work well with this pattern of internal company communications. Remember, switches were not computerized, and telephone lines were expensive. Large companies wanted many telephones: Yet, at any given time few of these telephones were connected to telephones located at a different location. The distributed PBX solved this problem nicely. The introduction of PBXs signaled a shift to a more distributed style of management for voice services.

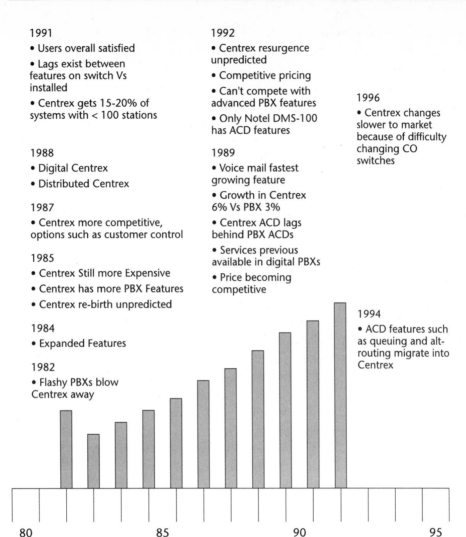

Figure 9.3   Centrex usage and history.

In the late 1950s, the next shift in management structure occurred — it was from a distributed to a more centralized type of architecture. The PBX turned out to be expensive in densely populated business districts, causing AT&T to build Central Office telephone switches that allowed PBX-type services for business users. This new Centrex service signaled a shift from a distributed to a more centralized style of management.

The seeds for the next shift in management structure began in the late 1960s, as regulation allowed vendors to compete with AT&T in providing equipment to users that connected to the PSTN. The relaxed regulations allowed competition, and technology advanced when PBXs became computer-controlled, giving vendors both the incentive and the ability to innovate. By the late 1970s, this new computerized PBX architecture, combined with the relaxed regulatory environment, promised advantages to users, and they started shifting to the distributed management structure of the PBX. Most believed that the combination of relaxed regulation and advancing technology made PBXs unbeatable.

This case focuses on the next shift in management structure from a distributed model, based on PBXs located at the customer site, to Centrex. It started in 1983, when the number of new Centrex lines bottomed out [14], as Figure 9.4(a) shows. This indicates that Centrex started to meet more needs of users. The increased success of Centrex as compared to PBXs is highlighted in Figure 9.4(b), showing the percentage growth per year. Users again found Centrex competitive to PBXs in terms of desired features and showed this preference by starting to buy more lines in 1983 and continuing to do so. At first, the shift was slow and subtle; it was not until 1985 that the total number of Centrex lines installed started to increase [14]. This shift back to Centrex was completely unpredicted because it was counter to the current belief that distributed solutions located at the customer site and managed by the user provided better service [7][14][34]. The shift continued into the mid 90s, with more advanced features such as ACD and VM migrating into Centrex service. The evidence that follows shows with certainty that a shift in management structure started in the early 80s and continued for more than 10 years.

As Centrex growth continued, the total number of Centrex lines installed started to increase in 1985, as Figure 9.5(a) illustrates. Figure 9.5(b) highlights this change by showing the slope of the graphs in Figure 9.5(a). This figure shows the installed base of Centrex lines compared to PBX shipments. The data for the two sets of PBX shipment numbers comes from BCR [14] and Verizon[4].

---

[4] A spreadsheet from Shelley Frazier from Verizon.

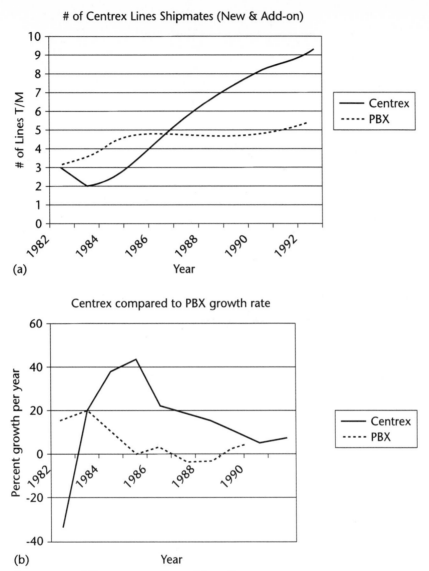

Figure 9.4   Shift to Centrex (lines shipped).

The data illustrates that the trend started in 1983 and continued well into the 90s, highlighting the strength of this shift in management structure from distributed (PBXs) to centralized (Centrex).

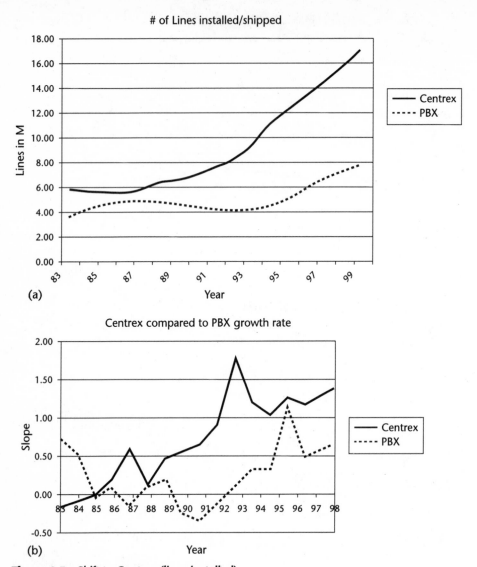

Figure 9.5 Shift to Centrex (lines installed).

Analysis of Management Shifts

As illustrated previously, voice services have seen several shifts in management structure. The conditions causing each of the shifts are unique to the environment at the time. Many factors influence the choice of management structure in voice services, such as regulation, technology, and market uncertainty. While complex, the evidence suggests that market uncertainty is correlated to the adoption of a centralized or distributed management structure, and in the last shift to centralized management, market uncertainty is the likely catalyst of this change. The many shifts of management structure over the history of voice services and the level of market uncertainty during the shifts are consistent with the theories in this book.

One important observation is that when market uncertainty is high, centralized management structure is not the architecture adopted by most users. Figure 9.6 illustrates this by demonstrating the correlation between date, market uncertainty, direction of change of market uncertainty, and popular management structure. Several important points in the history of voice services related to management structure are depicted, including the invention of the PBX, the invention of Centrex, *Carterfone* — the court case allowing non-AT&T solutions — the computer-controlled PBX, and finally the commoditization of PBX-type services. The arrows indicate the effect of the event on the market uncertainty and on the choice of management structure that is growing fastest. As shown, market uncertainty matters — centralized management grows faster than distributed management structure only when market uncertainty is decreasing.

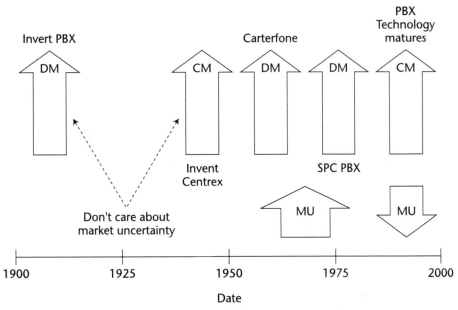

Figure 9.6   Analysis of market uncertainty and management structure.

The invention of the PBX made distributed management of voice services possible, yet that was not why it was invented. It was the calling pattern and the economics of the era that led to its development. Users at this time had only one choice of vendor for either PBX hardware or voice services. The incentive to innovate was absent because the law prohibited solutions other than the current AT&T offering.

Centrex was invented for the same reason the PBX was — the monopolistic telephone company decided it made economic sense. Again, users had little choice — they could buy voice services from AT&T via Centrex, or they could buy PBXs from AT&T. Product differentiation did not exist. Again, the incentive to innovate did not exist because users were locked into AT&T by regulation.

Before the *Carterfone* decision, market uncertainty was not a factor in the value of either the centralized or the distributed management structure. Users had only a single vendor — AT&T — because the law disallowed competition. Users had little choice in services and equipment — in the context of Porter's five-forces model, users had no power. As pointed out in Part One, the value of market uncertainty is realized only when users have choices, and this value is greatest when the choices come from many different service providers and when market uncertainty is high. This implies that before *Carterfone*, market uncertainty was not a factor in management shifts.

Finally, in the late 1960s, regulators provided the incentive for innovation. The *Carterfone* case opened up the voice service and hardware market — AT&T could no longer dictate users' choices of hardware or services. This development allowed market uncertainty to become an important factor in the choice of management structure. Now that users had choices, market uncertainty increased the value of this choice. Having many different PBX vendors created competition, giving an incentive for innovation.

Although *Carterfone* enabled competition, it was not easy for PBX vendors to experiment with new features because of the mechanical design of a PBX, but when PBXs adopted computer control as the dominant architecture, vendors suddenly had the capacity for easy and inexpensive implementation of new PBX features. This development created an environment in which it was easy to invent and test out new PBX features because the computer-controlled PBX requires only a software upgrade to add features. This event tremendously increased the market uncertainty as most new technologies do — it changed the landscape of possible applications. Neither PBX vendors nor users understood what worked best; there was lots of experimentation. As discussed previously, pundits believed that the distributed architecture for voice services via a PBX provided the best answer for most organizations. The flexibility of this new computerized PBX won the hearts of users and industry experts because of both the new powerful features and the control users now had with the distributed nature of PBX management.

The focus of this case is on this last shift in voice services, illustrated in Figure 9.4. It is the shift in the mid 1980s to Centrex that occurred after the technology of the SPC PBX matured that caused PBXs to become commodities. The availability of evidence from this shift is the strongest, and the regulatory environment allows the competition needed to make market uncertainty a factor in the management choice of voice services. The evidence shows that market uncertainty is the likely cause of this shift. To demonstrate this, the next section argues that market uncertainty decreased at the correct time and that other causes were not likely to have been the catalyst for the shift to the more centralized structure of Centrex. This implies that market uncertainty was the critical factor that caused management structure in voice services to shift to a more centralized model.

Why Users Migrated Back to Centrex

The shift of management structure in voice services in the mid 1980s is clear, but the cause is not. Did shrinking market uncertainty cause the shift to a centralized management structure, or was it something else? How big a factor was regulation or changing technology? To convince the reader that market uncertainty triggered this shift, the decrease in market uncertainty at the correct time is explained; next, other factors are ruled out as possible catalysts. This section demonstrates that a reduction in market uncertainty is the most likely cause behind the observed shift to a more centralized management structure.

Decrease of Uncertainty in Basic Voice Services

The evidence illustrates that market uncertainty began to decrease around 1983 for basic voice services and continued to do so well into the 1990s with advanced voice services. Many factors indicate decreasing market uncertainty in basic services: the convergence of basic features offered by different PBX vendors, the accuracy of predictions about the PBX market, the stable technology, and the investment in Centrex technology by vendors.

Comparing the feature sets of PBX vendors and Centrex services is one way to determine market uncertainty, as shown in Table 9.1. It contains a few features comparing the two leading PBX vendors (AT&T and Nortel), Centrex, and two new IP-PBXs (most of this data is from Alan Sulkin's yearly PBX report published by *Business Communications Review*). The two traditional PBXs have virtually the same feature set, and Centrex closely follows both of them. This indicates low market uncertainty by demonstrating that vendors have zeroed in on a feature set that meets most users' needs.

Table 9.1   Comparing PBX Features

VENDOR FEATURE	AT&T	NORTEL	CISCO	3COM	CENTREX
Call coverage	y	y	y	y	y
Call forwarding (off-premises)	y	y	y	y	y
Call forwarding (Busy/no answer)	y	y	y	y	y
Call forwarding all calls	y	y	y	y	y
Call forwarding – Followme	y	y	y	y	y
Call hold	y	y	y	y	y
Call park	y	y	y	y	y
Call pickup	y	y	y	y	y
Call transfer	y	y	y	y	y
Call waiting	y	y	y	y	y
Consultation hold	y	y	y	y	y
Distinctive ring	y	y	y	y	y
Do not disturb	y	y	n	y	y
Emergency attend access	y	y	y	y	y
Executive busy override	y	y	n	n	y
Facility busy indication	y	y	y	y	y
Incoming call display	y	y	y	y	y
Last number dialed	y	y	y	y	y
Loudspeaker paging	y	y	y	y	y
Malicious call trace	y	y	y	y	y

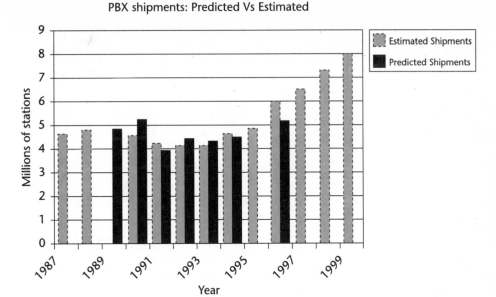

Figure 9.7   PBX market predictions.

One way to estimate market uncertainty is to examine how well predictions about the market match the actual market. Alan Sulkin, a well-known PBX expert, has published a set of predictions of the PBX market and later estimates of what the actual number turned out to be, in *Business Communications Review* each January. Figure 9.7 shows how well the estimates match the predictions, showing low market uncertainty from the late 80s through the mid-1990s. Before this point there is only secondary evidence of the accuracy of predictions; this data illustrates that the predictions were off, but only because of the unpredicted resurgence of Centrex, which shows a misunderstanding of the distribution of the market, but not its size. Thus, because the early 80s' predictions about the size of the PBX market were good, this indicates low market uncertainty.

In 1983, the PBX had a stable technology. The programmable PBX was the dominant design, and digital architectures were becoming common [3][4]. At first features exploded [13], but they soon settled down, with most vendors offering similar features [5][20][22]. The PBX was becoming a commodity due to the stability of the technology [5][19]. Future evidence showing a decrease of market uncertainty is the solidification of the secondary PBX market in 1988, again highlighting the commodity nature of PBXs [54]. The most important attributes of the PBX were becoming price and service, not distinctive features. By 1983, the SPC PBX was a stable dominant technology, illustrating that market uncertainty was decreasing.

One benefit of regulation was competition, which provided an incentive for vendors making Central Office switches to invest in creating new features. With the MFJ, the RBOCs became new potential customers for switch vendors such as Nortel and Siemens; this was a tremendous opportunity to break into the U.S. market[5]. This sort of commitment from vendors willing to invest in creating new features shows a reduction in market uncertainty.

Further evidence of low market uncertainty and less experimental activity is the lack of new venture capital (VC) funding in the PBX market [55] starting in the late 1980s. More evidence of lack of interest in new PBX products is their disappearance from trade shows at the beginning of the 1990s [56]. The inability to attract new VC funding for traditional PBXs and their invisibility at trade shows illustrates the low level of experimentation and innovation in PBX as the 1990s started, highlighting the low market uncertainty.

This illustrates that market uncertainty decreased at the correct time to explain the shift to central management in the early 1980s. It is convincing because several methodologies are used to measure market uncertainty, and they agree that it began decreasing at the correct time to explain this shift to Centrex service.

Decreased Market Uncertainty with Advanced Features

Market uncertainty decreased in the ACD market as it did in the general PBX market, but somewhat later. By the early 80s, PBX vendors knew it was important to offer ACD features [27]. Market uncertainty was still moderately high because the technology was new and complex. Initially, this caused a slow migration of ACD features into Centrex. This is evidenced by the fact that only Nortel offered a CO switch with ACD features [47][57] at the time. By 1991, ACDs existed even on low-end PBXs and KEY systems [47], indicating a decrease in market uncertainty as the technology matured. As with basic voice features, ACD features became commodity-like, again showing a reduction in market uncertainty. Illustrating this point is the convergence of features between standalone and

---

[5] From a conversation with my advisor (H. T. Kung), who remembered this at Nortel.

PBX-integrated ACDs [25][58]. Other evidence showing a decrease in market uncertainty is the narrowing of the price differential between standalone and integrated units [59], as shown in Figure 9.8. As expected, as features converged, so did price. Market uncertainty decreased with ACD features as users and vendors gained experience with them, similar to the evolution of more basic PBX services.

VM is another advanced PBX application. VM is less complex than ACD features, so customers and vendors understood the technology faster. One example of the reduction in market uncertainty in VM is the accuracy of market estimates by 1988 [33], as shown in Figure 9.2. Another measure of decreased market uncertainty is the similarity of VM features in PBXs [60]. Like ACD features, as VM technology matured, market uncertainty decreased.

Today, things are changing because data and voice are converging. One example is the birth of IP-based PBXs. IP-PBX vendors are implementing the important traditional features, as Table 9.1 illustrates, but they are mostly concentrating their research/marketing efforts on creating features that differentiate their products from others, such as auto-configurations of Internet phones. This illustrates that, in this new market of PBXs based on open architecture, the market uncertainty is high. This can be seen in Table 9.2 in the comparison of IP/PBX features. Chapter 10 in Part Three discusses Voice-over IP in more detail.

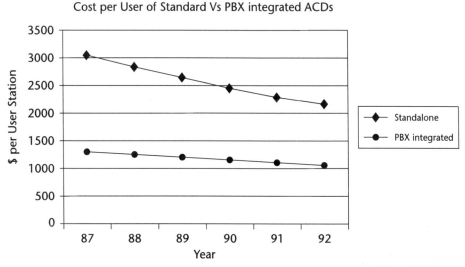

Figure 9.8   ACD price comparison.

Table 9.2 Feature Comparison between IP PBXs

VENDOR FEATURE	CISCO	3COM
IP features	y	y
Auto call-by-call BW selection	y	y
Auto phone install configuration	y	
Auto phone moves	y	
Configuration per phone (date/time)	y	
Configuration per phone (time zone)	y	
Event viewer	y	
Performance monitor	y	
E911 database configuration	y	
External SMDI interface	y	y
File transfer	y	
H.323 support	y	
Set IP precedence bit	y	
IP routable	y	
PRI protocol	y	
Redundancy	y	y
TAPI support	y	y
Web administration interface	y	
Web system documentation	y	
Web system speed dial		y
Whiteboard support		y
CTI services per user		y
CTI client per user		y
Unified messaging		y
Unified messaging client per user		y
Call retail recording		y
Simultaneous voice/data		y
LAN		y
Web-based training		y

Table 9.2 *(Continued)*

VENDOR FEATURE	CISCO	3COM
Attendant console		y
Hands-free answering on intercom		y
Paging services via phone		y
Telecommuter solution		y
KEY system		y
Open interface for phones		y
QoS		y
Remote administration		y
Call ID on analog lines		y
T1		y
Tie trunk	y	y
DID	y	y

Analysis of Other Factors

From the preceding text, it is clear that market uncertainty began changing at the correct time as PBX technology matured, but what other factors might be responsible for a shift in management structure? Such factors include technology and regulation. We show why the evidence suggests that these other causes were not catalytic in the migration of users to a more centralized management structure.

Voice services live in a complex environment because of changing technology, numerous regulations, and changing price structures. Similar to many computer applications, the rapid pace of technology change has affected the nature of PBX architecture, but this did not happen in 1983. Regulation did profoundly change the landscape of the telecom industry, but, as shown in the text that follows, it cannot explain the shift that started in 1983. The price of Centrex service and the structure of the pricing model have changed because of changes in regulation, but this cannot account for what happened in 1983. The possible causes of the shift to Centrex (technology, regulations, and price) seem unlikely candidates for this change in management structure.

Changing technology does not seem to have caused this management shift. PBX technology experienced its last major change in the mid to late 1970s with the switch to SPC architecture and digital voice [3]. AT&T and InteCom introduced the first digital PBXs [4] in 1980. After this fundamental paradigm shift, the technology changed only incrementally over the next 20 years. Mostly, as processor speeds grew, the capacity of PBXs greatly increased. Centrex technology was even more stable than PBX technology because the SPC architecture used in PBXs came from the CO switch[6]. In 1983, there was nothing changing in technology that could have caused the shift to Centrex.

Changing regulation does not seem to be the catalytic factor in the shift to Centrex in the mid 1980s. Regulation of voice services by the FCC was the price AT&T paid to be the only provider of those services for many years. This regulation dramatically affected the market for voice services, changing both the shape of the market and the players. At first, before the *Carterfone* decision in 1968, only AT&T was allowed to provided equipment that connected to the PSTN, which greatly limited customer choice. This landmark decision created a whole new industry, the Customer Premise Equipment (CPE) sector. This regulation created the environment for innovation by vendors other than AT&T; no longer did AT&T decide what choices its customers had [1]. This regulation did not cause the observed shift to more centralized management because it occurred too early.

In 1983, big things were in the works: The inconceivable was happening with the split of AT&T. AT&T, the courts, and the FCC agreed in 1984 to the Modification of the Final Judgment (MFJ), paving the road for future telecom deregulation and competition [2]. The breakup created the RBOCs, a group of competitive regional phone companies that could profit handsomely from Centrex. The RBOCs did not have to decide between marketing PBXs or Centrex because they were not CPE vendors. At the time, Centrex was considered dead, and this was not expected to change with the MFJ [14]. Most believed the MFJ would foster competition in the PBX market [14], and it did. The increase in PBX vendors and more venture capital illustrates this. Unexpectedly, this also helped Centrex because after the split-up, Centrex providers learned from the successes of PBX vendors about what features users wanted, and the RBOCs were in the perfect position to profit by providing them. These regulations occurred too late, though, to cause the shift of management structure that started in 1983.

---

[6] Personal email from Alan Sulkin.

The RBOCs viewed Centrex as a way to improve their bottom line and increased their marketing efforts. This marketing only helped the already resurging Centrex service. These new Centrex features had already succeeded in PBXs. With the FCC regulation [7] in 1985, the RBOCs were allowed to provide services such as speed dialing as Centrex services, and they already knew users wanted these features because they had learned this from PBX vendors. The proof is that these services were already successful in the PBX market and regulation existed to protect the CPE vendors from the monopoly the RBOCs still had on the local loop by restricting these advanced services. The creation of the RBOCs unexpectedly helped to revive Centrex services because it gave the RBOCs a profit incentive. This new marketing effort did not trigger the shift, though, because it promoted only those features already successful in PBXs, which helped exacerbate the already occurring shift in users' preferences for the more centralized management structure.

The cost of Centrex is now competitive with the cost of PBXs, but the comparison is complex. Historically, Centrex service was more expensive than owning a PBX. At first, Centrex was tariffed [7] by the FCC, giving little pricing flexibility to the RBOCs. Later, in the early 90s, regulators allowed the RBOCs to provide Centrex as a non-tariffed offering [47][53], giving the RBOCs what they needed to be more competitive with PBXs. A detailed comparison of the cost of owning a PBX as compared to buying a monthly service is given in [53], showing the competitive nature of Centrex pricing in the late 80s. It is hard to say when Centrex became competitive to PBXs, but it is now, and for some users Centrex may even be less expensive.

Comparing the cost of renting Centrex to buying and maintaining a PBX is complex because of factors that are hard to quantify. For example, what is the cost of office space to house a PBX? It depends on a number of assumptions. If the PBX is placed in a back closet[7], then the billing rate per foot is not clear because it should be less than the average cost per foot of the entire office space, but how much less? Another example is the value of having control of upgrade scheduling. Different companies will place different values on this control. Some businesses with sophisticated customer service requirements depend on advanced features and place a high value on user control, while others do not. Today, the cost of PBX ownership is comparable to the cost of renting Centrex service, but the comparison is complex, and different companies will arrive at different decisions based on their particular situation.

---

[7] Personal conversation with Leo Goeller.

From the preceding, it seems unlikely that other factors such as regulation, technology, or price caused users to begin favoring the centralized structure of Centrex over the distributed architecture of PBXs.

Market Uncertainty Caused This Shift in Management Structure

Factors other than market uncertainty don't explain the shift to Centrex that started around 1983; however, the reduction in market uncertainty can. It happened at the correct time, and nothing else seems likely to have triggered the migration of users to centrally managed voice services.

Neither technology change, nor regulation, nor cost is able to explain this shift from the distributed management structure of PBXs to the centralized architecture of Centrex service. Technology was not changing in any way that could have caused this shift. Regulation, while helping exacerbate the shift, cannot account for the timing of the shift in 1983 or the fact that Centrex service providers gave users only what they had become accustomed to from PBX vendors. Historically, Centrex has been more expensive than PBX, but it is now competitive. The decrease has been slow without crossing any obvious thresholds; even today comparing costs is difficult. None of the preceding factors can account for the shift observed in the voice market, but the shift in market uncertainty can.

The evidence from advanced PBX services also points toward market uncertainty as the major factor causing the management shift to the more centralized structure for these sophisticated features. Vendors view both ACD and VM features as *advanced applications*, but regulators see them very differently. ACD is a basic unregulated service, but VM is a store-and-forward application, subject to regulation by Computer II [7]. Disallowed by the FCC, it was not legal to provide Centrex voice mail until 1988 [34]. The examples of ACDs and VM show how market uncertainty plays an important role in determining when services migrate to a more centralized management structure. ACDs had no regulatory constraints and VM features did, yet both services had (and still have) most innovation at the edges, with the successful features migrating inside the network. Furthermore, ACD features that seem more natural inside the PSTN and had no regulatory barriers migrated more slowly than VM features to Centrex services.

Combining the technical and regulatory issues, the argument for migration based on market uncertainty is strong. VM has a strong technical argument for end-based implementations, and regulation prohibited the RBOCs from providing it as a Centrex service until 1988. On the other hand, ACD features have a good technical argument for centralized structure, and no regulations are pushing for end-based implementations. Yet, VM had a quicker adoption as a Centrex service — as soon as regulation allowed, VM successfully migrated into the network. Market uncertainty is the only factor that can explain what happened with both applications. In both cases, the centralized management architecture worked best after experimentation and education lowered market uncertainty.

VM services migrated faster than ACD, even given that their respective markets were of similar size, as shown in Figure 9.2 (VM data [31][33][61], ACD data [59]). Evidence of this is VM's quick migration into Centrex service as soon as regulation allowed in 1988 and the support of voice mail interfaces for both AT&T's 1AESS and Nortel's DMS-100 in 1988 [8]. The evidence of slower migration for ACD features is that in 1989, only the Nortel switch offered Centrex ACD features, while AT&T was still working on adding these features [34]. In 1989 there were about 4,000 lines installed with ACD features, compared to more than 1 million Centrex lines with voice mail. This shows that it took more time to figure out the correct feature set for ACDs. Even in the early 90s, only Nortel offered a CO switch with Centrex ACD tariffed service [47]. At the time, 15 percent of the installed Centrex lines had VM service [10] only one year after regulation allowed the RBOCs to provide store-and-forward voice services.

Regulation kept VM service outside the network, yet, once allowed, this advanced feature quickly became successful. Non-regulated ACD features could have migrated any time, but market uncertainty dictated a slow migration. Regulation alone cannot explain why VM migrated so quickly to a centralized structure and why ACD features migrated more slowly, but this theory based on market uncertainty does.

Looking at how basic and advanced PBX features started out with a distributed management structure, and then migrated to a more centralized structure, demonstrates the link between market uncertainty and management structure.

Conclusion

This chapter ends Part Two — The Case Studies. The evolution of voice services matches the theory presented in this book well. PBXs have a distributed management structure that permits easy experimentation, while Centrex has a central management structure that allows efficient implementation. PBXs incubate successful services, while Centrex service providers learn from PBX vendors about features that meet the needs of the *average* user.

The evidence showing the shift to the centralized management structure that started in 1983 and continued strong into the mid 90s is irrefutable. Nobody believed Centrex, once near death, could again become competitive with PBXs for providing voice services, but it did. Centrex's resurrection was completely unexpected, catching industry reporters, forecasters, and certainly PBX vendors by surprise. The theory based on market uncertainty, however, predicts this surprising resurgence of the Centrex market.

The next chapter starts Part Three — Applying This New Theory, linking market uncertainty to choice of management structure. It explores how to apply this theory to situations in today's complex environment. Applications such as Voice over IP, wireless services, and Web-based applications have extreme uncertainty about what services users will adopt with these new technologies. Industry experts and vendors do not know what services with what features will become popular. Massive experimentation is needed to meet these uncertain needs. It is important that management structure allows the experimentation required to meet this uncertain market. The ideas presented here will help chart the evolutionary path that these new applications will follow.

PART Three

Applying This New Theory

In this part, theories in this book are applied to current technologies in the hope of better understanding how to build successful network infrastructure and applications with technologies such as Voice-Over-IP (VoIP), wireless link layer protocols, and web applications and services. Part One explained the theory linking market uncertainty to management structure. It demonstrated that high market uncertainty favors distributed (end-2-end) management structure because it allows easy experimentation, and centralized management architecture works well when market uncertainty is low because it is efficient. Next, in Part Two, case studies of the voice and email markets illustrated how these technologies evolved as my model predicts. When market uncertainty was high, PBXs and distributed email succeeded; however, as market uncertainty decreased, users shifted to more centralized management structure (specifically Centrex and large, centralized web-based email services). The similarities between the evolution of services in the PSTN and Internet email provide strong evidence that this theory is correct. Moreover, the fact that these are two different services built on networks with very different infrastructure illustrates the generality of the theory. This last part examines new technologies that might change how we communicate, and eventually, how we live. An approach to maximize value of these yet-unknown applications is presented.

The first chapter in this part discusses sending voice over IP-based packet networks. For many years, large phone companies have discussed advantages of integrating data and voice into one network. They believed (hoped) that circuit-switching network infrastructure would allow this convergence of voice and data. Unfortunately for them, and for-tunately for users, it turned out that packet-switching infrastructure is the architecture being adopted by users to allow this integration. As discussed later, there are choices that must be made to build an IP packet-based infra-structure—should the current architecture of the PSTN be mapped into a similar scheme within the Internet, or should new, innovative architecture be attempted? Our choices today will profoundly affect the future.

Chapter 11 is about the wireless revolution that is here. Nobody seems to know how this infrastructure will unfold or what services will be prof-itable. There are different models for providing broadband wireless Inter-net services: N^{th} Generation, (3G for now) architecture with its carrier-based centralized management structure, and 802.11 with flexibility of choice in management structure. For now, 802.11 is growing from the bottom up and hot spots are sprouting up at alarming rates. While many business models view these as competing wireless technologies, this book does not; rather, I propose a model of cooperation rather than competition. In this world, users pick the best Internet connectivity from the available options. This model illustrates how each technology gains value from the other, which portends a friendly coexistence of both technologies.

Chapter 12 in this part discusses web applications and web services. Some web-based applications (for example, Hotmail for email, MapQuest for driving directions, and services such as CNN.com) are very successful, becoming part of our everyday lives. Web services are based on ideas from web applications and have the potential to transform the way businesses interact internally, and with outside partners. Both web-based services and applications share attributes of both centralized and distributed manage-ment architecture. The success of web applications and anticipated success of web services fit well with predictions of theories from Part One.

CHAPTER 10

Voice-over IP: SIP and Megaco

This chapter examines the technology of Voice-over IP (VoIP), which sends voice-over IP packet networks. It looks at two different proposals — Session Initiation Protocol (SIP) and megaco/H.248 — in the context of their management structure. Using theories from Part One, SIP and megaco/H.248 are compared, illustrating that one advantage SIP has over megaco/H.248 is that it allows a choice between a distributed and a centralized management structure. This flexibility has the greatest value when market uncertainty cycles between high and low. I discuss why SIP is a clear winner over megaco/H.248 for certain markets because megaco/H.248 locks users into a centralized management architecture, which makes it difficult for users to experiment. When market uncertainty is high, forcing users into centralized management stifles the innovation needed to meet users' needs. Because SIP offers the advantages of megaco/H.248's centralized structure but still allows users to innovate, my theory predicts that SIP will succeed in more situations than megaco/H.248 even though megaco/H.248 offers a more reliable billing structure for service providers.

VoIP

The economics of sending voice over packet networks, coupled with the technology that ensures acceptable quality of service, is causing the convergence of data and voice. It is unclear, though, what VoIP management architecture will best meet user needs, what services will be most successful, and at what location in the network these services should reside. This chapter links the value of VoIP-based services to the management structure of the service and the market uncertainty. It illustrates how high market uncertainty increases the value of the flexibility of SIP, because it gives users choice in management structure.

Historically (as discussed in Chapter 3), the end-2-end principle has been important in designing Internet protocols and the applications that use them. One proposed architecture for VoIP is the SIP [1] protocol, which is supported by the IETF. It promotes user innovation because it allows end-2-end operation. It also offers the advantage of central control via a proxy server. Another architecture for providing VoIP is megaco/H.248 [2], which was developed by the IETF and ITU. Megaco/H.248 is a centralized architecture similar to the current PSTN and does not allow true end-2-end use. The ideas in this book illustrate why SIP has a higher expected value when market uncertainty exists because it offers the advantages of end-2-end services when innovation is important, along with the advantages of centralized control and management when efficiency is paramount, at the user's choice.

VoIP means different things to different people. It might describe how traditional dumb telephones talk to a megaco gateway in order to convert the analog voice into IP packets, or it might explain how smart SIP phones establish connections to each other, or it might define the protocol for IP phones to become a Media Gateway controlled by a Media Gateway controller, as described in RFC3054. This chapter focuses on the market for new IP-based voice infrastructure. The question addressed in this chapter is the estimation of value between different architectures for a new IP-voice system. Will forced centralized control, along with its implied efficiency and the robustness of the megaco/H.248 architecture [3], prove more valuable than the flexibility inherent with the SIP model? This chapter discusses the potential value of the market for new IP-voice systems.

The question of where to place services in IP telephone networks is complex [4] and depends on many variables, including the capabilities of the end systems, the amount of interaction with the end user, and the architecture of the network infrastructure. Networks with dumb devices, such as the PSTN, force services to be implemented within the network because of the limited ability of the end devices. Networks having smart end systems allow more choice in where to locate a service, but other factors affect the location of services in these networks. Services dealing with how to reroute a call to a device that is not working require network intervention because smart devices not connected to the network can't participate in any service offering. Sometimes, the user of the end device is needed to help decide what to do — for example, transferring a call. Assistance from the end device might be helpful to the user, for example, when user interaction is desirable such as handling a new incoming call when the user is already busy with at least one other call. In this case, the user may wish to know who the incoming caller is, and maybe the topic of the call, and decide if the new call is more important than any of the current ones — involvement of the end device makes this much easer. Finally, services such as conference calls may or may not require network intervention. Flexibility in where to place services in IP telephone networks is valuable because it promotes innovation of services that meet user and service provider needs.

This chapter focuses on voice service over the Internet, which is a network allowing end-2-end services. Voice services built using the Internet can have centralized or distributed architecture. Conference calling implemented with SIP is a good example of such a service. It can be implemented with a distributed structure without the network providing any services — each end device manages the different audio streams. Alternatively, it can be implemented within the network using a centralized network server to mix the many audio streams [4]. The end-2-end implementation is more flexible, but it is not as scalable as the more centralized architecture.

This book's theory will highlight the intrinsic value of SIP because it allows users and small service providers to experiment with new services. SIP users can bypass any centralized management structure and create new services without anybody knowing about these services. I illustrate how SIP allows the best of both worlds: centralized management via proxy SIP servers and end-2-end services when they make sense. This is not the case with megaco/H.248, which forces centralized control with all services. I discuss how megaco/H.248 will coexist with SIP and how, in the long run, megaco/H.248 might disappear because SIP is able to provide the advantages of megaco's centralized structure while allowing the flexibility to locate services either within the network or at its edge.

SIP

SIP is designed for setting up multimedia connections between consenting parties. One of the many uses of SIP is to provide voice services over the Internet. SIP is similar to HTTP, having text headers that people can understand.

It is a flexible protocol because it allows true end-2-end applications as well as applications requiring centralized control via the use of SIP proxies. When SIP is used end-2-end, users can experiment by extending the protocol. Many wireless phone providers, such as Ericsson or Nortel, are adopting SIP because they are beginning to understand that service providers other than traditional phone companies will be providing some (maybe most) of the future applications for wireless phones and other wireless devices.

SIP is used in conjunction with other protocols, such as Session Description Protocol (SDP) and Real Time Protocol (RTP), allowing multimedia sessions to be initiated, carried out, and then terminated. SDP is the protocol used to describe the characteristics of the connection, such as the protocol used for data transfer (RTP/UDP), the type of media (voice), and the format of the media (that is, the voice encoding). It is needed so that both sides can exchange information about what the upcoming data stream will look like. RTP is the application protocol enabling data transport — it is unreliable, but it has sequence numbers so packets are delivered to applications in the order sent. Together with these other application protocols, SIP allows building Voice over IP applications on the Internet.

SIP is not a complex protocol. It consists of commands (messages) with headers that describe attributes of the command. To initiate a call in SIP, the INVITE message is sent along with a set of headers containing information required to establish the call. The example that follows is the end-2-end use of SIP — Alice is calling Bob directly without the help of an SIP proxy server. Here is an example of what the INVITE command (taken from RFC2543.bis09) might look like if Alice at Harvard is calling Bob at BU:

INVITE sip:bob@pcbob.bu.edu SIP/2.0

Via: SIP/2.0/UDP pcalice.harvard.edu;branch=z9hG4bK776asdhds

Max-Forwards: 70

To: Bob <sip:bob@pcbob.bu.edu>

From: Alice <sip:alice@pcalice.harvard.edu>;tag=1928301774

Call-ID: a84b4c76e66710@pcalice.harvard.edu

CSeq: 314159 INVITE

Contact: <sip:alice@pcalice.harvard.edu>

Content-Type: application/sdp

Content-Length: 142

The first line of this SIP message contains the INVITE message. This command is directly asking Bob at Boston University (bob@pcbob.bu.edu) if he would like to talk. The protocol used is SIP, version 2.0. Following this method is the minimum set of headers describing the required information for Bob to talk with Alice. These headers are described next:

Via.   This is the address (pcalice.Harvard.edu) where Alice is expecting to receive responses to this request. It also has a branch parameter to identify this transaction.

Max-Forwards.   This is the maximum number of hops this request can take.

To.   This indicates where the SIP message is going — directly to Bob at Boston University (bob@pcbob.bu.edu).

From.   This is the address of the calling party — Alice in this case (alice@pcalice.harvard.edu).

Call-ID.   This contains a globally unique identifier for this call.

Cseq.   This contains a sequence number for the particular method specified.

Contact.   This is the direct route to Alice so that Bob can directly communicate with her. In this case, it is the same as the "From header" because a proxy is not being used.

Content-Type.   This describes the content type of the message and comes from the MIME world; in this case, the message is a session description using the Session Description Protocol (SDP).

Content Length.   This is the length of the entire message with headers.

All the headers apply to the method specified in the first line of the SIP message.

In this particular end-2-end case, the message is going from Alice's system at Harvard to Bob's system at Boston University, without any SIP proxies involved in the transaction. This simple end-2-end use of SIP is illustrated in Figure 10.1. This end-2-end architecture illustrates how users directly communicate with each other without any server within the network knowing this is taking place. Next, each step of this simplified example is explained:

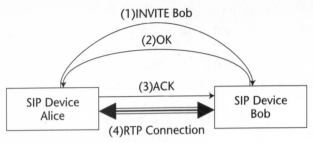

Figure 10.1 End-2-end SIP architecture.

1. Alice sends the INVITE message to Bob, asking him if he wishes to join her in a conversation.

2. Bob receives this message from Alice and responds with the OK message, indicating to Alice that he is glad to speak with her.

3. Alice receives Bob's OK message and sends him the ACK message to indicate the establishment of the session between them.

This is a good example of a pure end-2-end application because nothing within the network is aware of the application Alice or Bob is running. If Alice and Bob choose to experiment, they can because the network will be unaware of what they are doing; they are the only ones to be affected by their actions. For example, Bob and Alice can agree to experiment with new messages and/or new headers that will enable a service that they want to implement. With SIP, they can experiment between themselves with end-2-end SIP without modifying any servers within the network. Bob and Alice can change the SIP however they want with this distributed mode of operation because only they are aware of what they are doing.

SIP provides a more centralized management architecture by allowing end users to communicate though an SIP proxy. This proxy is responsible for establishing a connection between the two end users. The following is an example of an INVITE request where both end users communicate via an SIP proxy. Bob's proxy is sip.bu.edu, and Alice's proxy is sip.harvard.edu:

INVITE sip:**bob@sip.bu.edu** SIP/2.0

Via: SIP/2.0/UDP pcalice.harvard.edu;branch=z9hG4bK776asdhds

Max-Forwards: 70

To: Bob <sip:**bob@sip.bu.edu**>

From: Alice <sip:**alice@sip.harvard.edu**>;tag=1928301774

Call-ID: a84b4c76e66710@pcalice.harvard.edu

CSeq: 314159 INVITE

Contact: <sip:alice@pcalice.harvard.edu>

Content-Type: application/sdp

Content-Length: 142

This example illustrates the similarities with both end-2-end and proxy models. The differences are highlighted in bold. The INVITE method has Bob's proxy, not his direct address. The "To" header also has his proxy instead of his direct address. Finally, the "From" header has Alice's proxy address, instead of her direct address. These are the only differences.

This more centralized structure for SIP is illustrated in Figure 10.2, where SIP end users do not directly establish sessions, but instead communicate with an SIP proxy. This SIP proxy is a network server, making this structure not end-2-end. Here is a simplified example of Alice asking Bob to talk when they are both communicating via different SIP proxy servers:

1. Alice sends the INVITE message through her proxy server (sip. harvard.edu), asking Bob if he wants to talk to her.

2. Alice's proxy server forwards the INVITE message to Bob's proxy server (sip.bu.edu).

3. Bob's proxy server forwards the INVITE message to him — this is the first time he is aware that Alice is asking to establish a connection to him.

4. Bob sends the OK message back through his SIP proxy server.

5. Bob's proxy server forwards the OK message to Alice's proxy server.

6. Alice's proxy server sends this OK back.

7. Alice sends the ACK message to Bob directly, indicating session establishment. This is the first time that Alice has directly communicated with Bob.

One important architectural question arising from the preceding example is how much state an SIP proxy server keeps about each connection it sets up. SIP allows proxies that are stateless or stateful. A stateless proxy retains no knowledge of what the end users are doing; a stateful proxy keeps information about what all end users are doing. For my argument, the statefulness of the server does not matter. The fact that the proxy exists and must understand the session setup between the two communicating parties is the important factor that indicates a more centralized style of management structure.

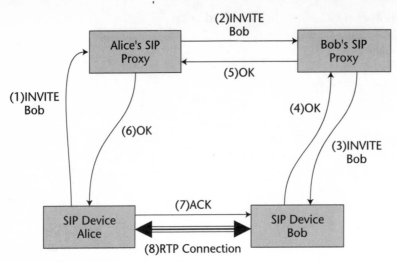

Figure 10.2   Centralized SIP architecture.

One important factor is that both of the preceding examples are identical as far as the end devices are concerned. As Figures 10.1 and 10.2 illustrate, the messages that Alice and Bob see are the same with or without the SIP proxy between the end users. The only differences are in the headers that indicate name/address information; the interchange of messages is the same in both end-2-end and proxy mode. This is important because it allows easy migration of services from the network's edge into its core.

SIP is a nice solution to VoIP, but it is not the model of choice for traditional centralized telephone service providers. Some traditional telephone companies believe in customer lock-in and worry that with SIP it is easy for users to switch service providers or do without a service provider altogether by adopting end-2-end SIP. By pointing to a different SIP proxy or directly connecting to another SIP user, SIP customers will have choices in how they use the SIP services. Some traditional carriers are afraid that with SIP, users will not choose them as their primary service provider.

Figures 10.1 and 10.2 show why SIP is the best of both worlds: It offers the advantages of end-2-end applications, or the advantages of centralized management using proxy servers, or a combination of the two. On one hand, SIP allows true end-2-end use, as Figure 10.1 depicts, which promotes innovation; on the other hand, SIP allows the business and technical advantages of centralized control by funneling the connection requests to an SIP proxy, as illustrated in Figure 10.2. SIP seems to give both users and administrators what they want because it allows experimentation when needed and then allows services to be placed where they fit best, and can most efficiently be deployed for many users, either inside the network, at the end device, or a combination of both.

Megaco /H.248

Megaco/H.248 is a protocol with a central management structure, jointly developed by the IETF and ITU. It is based on the traditional way voice services are provided in the PSTN. Inherently it is a less flexible architecture than SIP because it prohibits two end users from establishing communication directly. Users must request service from their Media Gateway. Unlike SIP, which can be operated in a peer-to-peer model, megaco/H.248 forces a master/slave relationship between the end device and its Media Gateway. A Media Gateway Controller (MGC) controls this Media Gateway (MG). MGCs communicate with each other to establish connections on behalf of their client MGs. As with the current PSTN, the megaco/H.248 architecture is centralized because a Media Gateway Controller must take part in any voice communication between any two users.

Figure 10.3 illustrates the forced centralized structure of megaco/H.248. It shows two types of end devices: The left side is a smart megaco IP phone (RFC3054), which acts as both end device and Media Gateway. The right side is a more traditional use of megaco/H.248 — connecting PSTN phones to the Internet. In this case, Bob has a traditional phone that is controlled by Bob's Media Gateway. Bob's Media Gateway is controlled by his Media Gateway Controller; Alice's smart megaco/H.248 IP phone is controlled by her Media Gateway. With megaco/H.248, when Alice wishes to call Bob, she must interact with her Media Gateway. The Media Gateway then relays this request to its Media Gateway Controller via the megaco/H.248 protocol. Finally, the Media Gateway Controllers coordinate the connection between Bob and Alice via SIP (or another protocol such as H.323). Media Gateway Controllers interact with each other to establish a connection between Alice and Bob; they maintain state information about this connection, which implies a centralized structure.

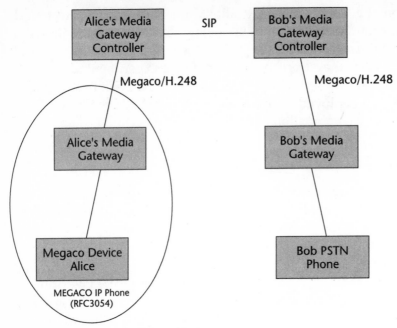

Figure 10.3   Megaco/H.248 architecture.

Megaco/H.248 is based on a proven robust set of protocols used in the PSTN to provide reliable voice service over a circuit-based network. Megaco/H.248 is the result of decomposing the pieces of a CO switch, such as Nortel's DMS100 or AT&T's 5-ESS [5], and building these components back together using the packet-switched paradigm of the Internet. This is not as far-fetched as it sounds because signaling in the PSTN is accomplished on the SS7 [6] network, which is also packet-based. Phone companies like this model because of its familiarity; they don't need to change their business model because the technology is similar to their current system.

The idea behind megaco/H.248 is to decompose a CO switch in the PSTN. Figure 10.4 is the interaction of Bob's dumb phone with his Media Gateway, illustrating some of the similarities of the PSTN and megaco/H.248, because both are based on the control of dumb end devices. This architecture makes it hard for users to experiment because it forces all services into the network where users can't experiment with them. Without access to servers within the network (specifically, the Media Gateway Controller), a user is unable to change the rigid procedure that sets up the call. If a user decides other information should be included in the call setup, he or she cannot include it without modifying a server within the network. Megaco will work because it is based on the successful and well-understood PSTN.

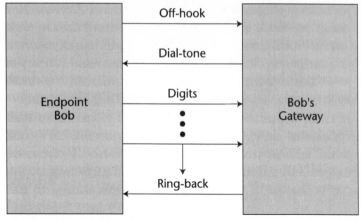

Figure 10.4   Dumb end points for megaco.

Applying the Model to SIP and Megaco/H.248 Architectures

This section applies the preceding theory to predict which of several different proposals for Voice over IP infrastructure will best match user needs in the market for new systems with IP-capable phones. Session Initiation Protocol (SIP) [1] and megaco/H.248 [2], developed between the ITU and the IETF, represent two different management structures for providing VoIP. To the user of a *traditional IP phone*, dialing a phone number with either of these two protocols is indistinguishable. From the point of view of a service provider, these protocols are very different. Modeled after a CO switch, H.248 has a centralized management structure. The megaco/H.248 Media Gateway and it's controller for a particular end device must know what the end point is doing. In this model, the network has intelligence similar to that of the PSTN; the centralized management structure is forced by the protocol. SIP is more versatile than H.248 because it allows, but does not force, a centralized management structure. SIP allows a continuum of architectures: from end-2-end, where all services reside in the IP-phone, to completely centralized, where all services are network provided, as well as a middle ground, where services are built wherever they fit best.

Predictions Based on the Theory

From this theory, if market uncertainty cycles between high and low as expected, then SIP is expected to better meet users' needs because it allows the trial and error required to find what services are the most valuable and to determine what feature set these services should have. When market uncertainty is high, the ability of applications to have the end-2-end structure allowed by SIP encourages the experimentation needed to figure out what services users want, with what feature set. With SIP, when market uncertainty is low, services can be implemented in the best technical way: with no assistance from the network, with no assistance from the end-device, or with the network and the end device working together. For now, the market uncertainty surrounding services to be provided with VoIP is very high, which implies the great value of SIP's flexibility in choice of management structure. One piece of evidence for this is the wide variance of features between IP-PBXs of different vendors (see Chapter 9), compared to the similarities of features from vendors of traditional PBXs. Market uncertainty is high now, which implies that the flexibility that SIP allows in the choice of management structure is highly valued.

Central management of VoIP services makes sense in many cases because of its efficiency and its ability to control and police the network. It is also important because some services either require it, such as call handling when the called device is unable to help, or benefit from it, such as large conference calls using a server to mix the many audio streams. SIP allows those services requiring network intervention to have it via a proxy server. This dual nature of SIP is well suited to foster the most successful services in the context of user adoption. Users can develop new services with end-2-end structure that, when successful and advantageous to centralized structure, can be seamlessly integrated into a more efficient centralized SIP model with an SIP proxy. Because end-2-end and centralized models of SIP are transparent to the end users as far as the protocol is concerned, the migration of chosen services from end-2-end architecture to a centralized mode is not as hard because the end points don't need to change. SIP seems to offer the best possible situation: easy innovation, efficient management, and ability of services to migrate from end-2-end services into a centralized architecture.

As discussed in the Foreword to this chapter and in [4], different services have technical reasons to reside at different spots. Services that find user interaction helpful should have it. On the other hand, services that require action when the end device is unreachable require network assistance. Services such as conference calling can have either structure: End devices or a central server can mix the different audio streams. SIP allows all these scenarios.

In contrast to SIP, megaco/H.248 requires centralized control with all services. It does not allow the end device to help provide a service because it assumes this device is too dumb to participate in providing service. This forced centralized structure is bad for two reasons. First, when market uncertainty is high, my theory illustrates the need for experimentation that is hard to do with megaco/H.248. Second, megaco/H.248 does not allow services to reside at the optimal point. Services that might take advantage of end-device intelligence can't do so because of megaco/H.248's structure. The centralized control mandated with megaco/H.248 is of less value than the flexibility of SIP when market uncertainty is high or changing.

The big phone companies are champions of megaco/H.248, and they will provide services with this architecture that forces a centralized management structure because this is what they know and believe in. It will be popular at first because it will be competitive in price, be highly reliable, and offer good quality of service. If the big service providers using the megaco/H.248 model are smart, they will be looking constantly at the SIP world to find new services (such as the best feature set for conference calling) that can be ported to the centralized structure of megaco/H.248. This will keep the megaco/H.248 architecture alive for at least a while.

Believers in the megaco/H.248 architecture want to lock in the customer to a single service provider that controls the basic IP service as well as user access to advanced services. They want end devices locked into a particular Media Gateway or Media Gateway Controller. This security of the megaco/H.248 model is false; it assumes that users will not have choices. If megaco/H.248 succeeds, then it is likely that other service providers will enter the market and vendors will give users a choice of which network server to access. These vendors must please users, not just big service providers, and users have never liked vendor or service provider lock-in.

Phone companies will have some opportunity to lock in customers with SIP. With SIP, service providers can program the default SIP proxy into the end device. Many customers are unlikely to change this default because some users don't like to change anything with their electronic devices. This illustrates that SIP will offer service providers some customer lock-in, but not as much as megaco/H.248 provides.

For now, the efficiency of sending voice over a packet network is enough to stimulate service providers to build both SIP and megaco/H.248 infrastructures. Both megaco/H.248 and SIP architectures will coexist. In the end, SIP will prevail (in the market I am interested in) because it offers the best of both worlds with a single protocol. The economic advantage of a single protocol will cause SIP to be the dominant design for the new IP-based phone infrastructure. SIP will meet the needs of the big centralized service providers, the smaller service providers, and the needs of users who want to experiment and innovate.

Conclusion

In this chapter, two different architectures to provide VoIP are analyzed in the context of the theories in this book. The value of SIP over megaco/H.248 in the market for new IP-based infrastructures is demonstrated due to high market uncertainty in this market. The ability of SIP to operate in both the distributed end-2-end structure and the centralized proxy architecture provides the flexibility needed to capture value in this uncertain market. If market uncertainty cycles between high and low, as it has in traditional voice services, this flexibility with choice of management structure will provide the most value. Megaco/H.248 will work well in conditions of low market uncertainty; SIP will work well in all conditions.

The next chapter explores how the future of wireless infrastructure will unfold. Will 802.11 and 3G compete with each other, each hurting the other? Or will these two wireless technologies cooperate, each adding value to the other as they team up to provide seamless wireless broadband Internet/LAN access? This chapter discusses the technology and the management structure of both technologies. There are many similarities between SIP and 802.11 in the context of allowing flexible management structure. Similar to SIP, 802.11 infrastructure promotes distributed management, but it allows centralized structure. 3G technology is discussed in terms of what it offers that 802.11 can't and how innovations in 802.11 environments can create an option for 3G service providers. It also discusses other business models of centralized 802.11 management, such as the consolidation model that Boingo and Joltage are exploring. For now, nobody knows what the wireless infrastructure will look like 10 years down the road or what services users will adopt, but it is known that giving users choices about the management structure will be an important attribute of this future wireless infrastructure.

CHAPTER 11

Coexistence of 802.11 and 3G Cellular: Leaping the Garden Wall

This chapter uses the real options approach [1][2] described in this book to analyze the growing market for wireless network-based applications. This model illustrates how high market uncertainty increases the value of distributed 802.11 network infrastructure because it allows easy experimentation. It demonstrates that when market uncertainty is low, centrally managed next-generation cellular, or 802.11 wireless networks with centralized structure, can capture value from the innovations in 802.11 networks that have distributed management. I would like to thank my two coauthors of this chapter, Nalin Kulatilaka (Boston University) and Scott Bradner (Harvard University), for many good ideas and help with the writing.

The potential benefits of wireless connectivity are immense. Untethering physical connectivity can offer a myriad of opportunities for enterprises by allowing their current activities to be done more efficiently and making entirely new business activities possible. Wireless technologies will also make life more convenient for individual consumers. We are well on our way toward this wireless revolution. The penetration of cell phones has reached nearly half the U.S. adult population (and nearly all of the adult populations in some parts of Europe and Asia). Wi-Fi (named by the Wireless Ethernet Compatibility Alliance) wireless LANs are popping up in business facilities, university campuses, neighborhoods, and public places such as airports, hotels, and coffee shops. The future path, though, is unclear. There are several different proposals for how broadband wireless Internet connectivity should be provided. Will a centralized infrastructure, such as that planned for the next generation of cellular services (3G, 4G), or the more distributed structure of 802.11 LANs win, or will we live in a world where both technologies coexist, each creating value for the other? This chapter explains why cooperation between Wi-Fi and the next-generation cellular network infrastructure is the most likely to succeed because it creates the most value for both services.

The Walled Garden

The metaphor of a walled garden conjures up the dominant model for the next-generation cellular service, as illustrated in Figure 11.1. Traditional mobile telecommunications firms can be thought of as operating within a walled garden, where the applications available to the end users are chosen and, in some cases, even developed by the network operator. United States and European wireless carriers are looking to deploy their 3G networks within the familiar walled garden. [1] For instance, subscribers are offered a menu of choices (that the service provider determines) like call forwarding, voice mail, and *## services. Developed in a world of voice-centric communications, the walled garden approach was extended by wireless carriers in their early forays into data services using the Wireless Access Protocol (WAP). It is now widely believed that the failure of WAP was largely due to the unavailability of enough applications for the extra cost of the service to be seen as worthwhile to end users. The developers of WAP applications were constrained by having to rely on operators not only for accessing their services but also for providing the experimental environment. As a result, there was insufficient innovation for WAP to catch on and tip the market. The applications that emerged were too costly and did not have sufficient appeal in the eyes of the end users to justify their cost. In WAP's case, this walled garden proved too limiting to promote enough innovative applications. This strategy is unlike the unfolding landscape of 802.11 technology, which can be managed in a distributed fashion and which lets groups of end users or entrepreneurs experiment with potential applications that are vital to solve their business or personal needs — an *open garden* model.

[1] The exception to the rule is Japan's NTT-DoCoMo service where the business model clearly encourages innovation by third parties.

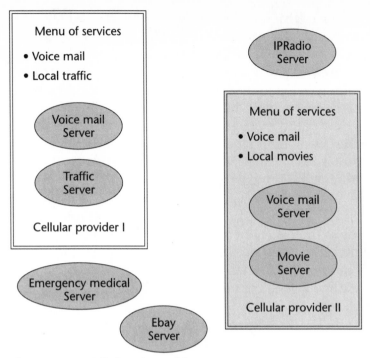

Figure 11.1   Walled garden architecture.

The early euphoria around 3G has since been taken over by widespread pessimism. There are grave concerns about whether sufficient applications will be available to generate the revenues needed to recoup the extremely large initial outlay for spectrum licenses and network expenditure. The architecture for providing broadband wireless network infrastructure of the next-generation cellular services is based on a centralized infrastructure. These cellular networks promise ubiquitous access and reasonable prices with bandwidth at the low end of broadband speeds. Similar to the PSTN, such a centralized structure of cellular wireless networks makes it hard for users to experiment on their own with new services. Yet, they provide efficient service over vast areas, making them important for the deployment of services once these networks have been developed and applications found that users are willing to pay for.

Meanwhile, there has been a groundswell of activity by individuals and enterprises deploying 802.11 wireless LAN networks. These networks are cropping up as home networks, enterprise campus networks, and hot spots around congestion points like airports, hotels, and coffee shops.

In each case, these networks give untethered high-bandwidth access to the Internet and Internet-based applications. Buoyed by the IEEE standard, both 802.11 network equipment and end-user devices are available at very low costs. Unlike 3G networks that use an expensive licensed spectrum, 802.11 networks run on a free public spectrum. Critical to this argument explaining why the *open garden* model has greater value to cellular carriers is that 802.11 networks can be managed in a distributed fashion. As a result, groups of end users or entrepreneurs can experiment with potential applications that are vital to solve their business or personal needs. It is my belief that this will result in fostering much greater innovation.

The success of a 3G strategy will be greatly enhanced if mobile operators offer the innovative applications that will result from entrepreneurial experimentation through their 3G networks. Vital to this success would be a pricing structure that allows the innovators to capture a fraction of the revenues, while 3G carriers could use their centralized management and billing services to capture a fraction of this revenue. This creates a win-win scenario for all the players.

It should be recognized that a large component of the value of wireless networks is derived from the options they create. The options inherent in the network, in effect, confer rights to offer new services. As with any option, the value of such networks will increase with increasing market uncertainty. The option value will also depend on the abundance of available options, such as innovative new services that can be launched from the network. By allowing for easy experimentation, the distributed 802.11 networks will foster greater innovation. As market uncertainty increases, so does the amount of innovation and the value of this innovation; hence, the option value of the network grows. The commercial deployment of wireless services, however, is best accomplished via centrally managed networks; a cooperative arrangement between 802.11 and 3G is the best of both worlds and will likely become the dominant solution.

Background on 802.11 to 3G Cellular

3G Cellular: Technology

Most of the world is currently using second-generation (2G) cellular systems. These systems are based on one of two dominant 2G technology standards: Time Division Multiple Access (TDMA) and Code Division Multiple Access (CDMA). More recently, some operators have upgraded their systems to an intermediate step (2.5G) as a stepping-stone toward the eventual offering of 3G services. The appropriate 2.5G network depends not only on the installed base of 2G but also on the planned technology strategy for the next generation. What will come after 3G? For now, there is extreme uncertainty about what the dominant technology will be and what services will be profitable for service providers.

The primary motivation to adopt 3G was to address the capacity constraints for voice services offered on existing 2G networks. The additional capacity is achieved via several methods. First, 3G cellular is more efficient in using bandwidth. Second, 3G networks will operate in higher frequencies, allowing for a much wider bandwidth capacity. The higher frequencies also necessitate smaller cells that lend themselves to greater spectrum reuse (in nonadjacent cells), which further enhances capacity.

Voice capacity increases may be only the tip of the iceberg for sources of value. The greatest benefits are expected from voice services that are enabled by packet-switched (and, hence, always on) data. The potential 3G data services could also leverage additional attributes of knowledge of location, presence, and greater intelligence in the network. Vendors of cell phones are building IP-enabled phones because they believe that services will emerge that users demand. Third-generation technology is a big advance over 2G for data applications because of the increased data rate and the always-on paradigm. 3G cellular networks are moving toward the convergence of voice and data networks seen elsewhere. Nobody knows what the landscape will look like when data and voice finally converge, and 3G cellular networks that allow VoIP will be better able to fit into whichever world evolves. It is these advantages of moving from 2G to 3G technology that are causing companies to build equipment and networks.

The evolution to 3G cellular services is complex because of the variety of 2G systems in the world including Global System for Mobile Communications (GSM), a TDMA technology used extensively in Europe and some parts of the United States, and IS-95, the popular CDMA system used in North America. Table 11.1 illustrates how these standards are expected to unfold, as described in [3][4].

Table 11.1   Evolution of Cellular Technology

GENERATION	TECHNOLOGY	STANDARDS
1G	Analog	AMPs
2G	Digital, circuit-switched	CDMA (IS-95), TDMA/GSM

(continues)

Table 11.1   Evolution of Cellular Technology *(Continued)*

GENERATION	TECHNOLOGY	STANDARDS
2.5G	Digital, circuit-switched for voice, packet-switched for data	CDMA (IS-95B), 1XRTT, GPRS
3G	Digital, circuit-switched for voice, packet-switched for data	CDMA2000, UMTS

3G Cellular: Management Structure

The different technologies of 3G are not critical to this argument, but the management architecture is. All of the preceding 3G technologies have a similar management structure because they force centralized management on the user. Current cellular network users have no input into how the network is managed, and generally they can't create new applications to experiment with. Users have little power over anything but their end devices, and they depend on a large service provider for everything. Service providers have agreements among one another that allow users to roam. This points to a centralized structure of next-generation cellular networks.

Cellular phone networks are mostly designed and managed by traditional telephone company types — bell heads, as some call them. It is not surprising that they are based on a centralized structure because that is what these providers know best. With current models of cellular services, end users can never directly connect to each other, but must follow a rigid set of protocols to have a circuit established between them by a centralized switch. Once the connection has been established, the network must manage the handoff if the user moves. This centralized architecture works well, just like the wired PSTN, but is not conducive to innovation, just like the wired PSTN, because of the stifling effects of centralized management on the ability of users to experiment.

A simplified cellular wireless network architecture is illustrated in Figure 11.2. It is very basic, containing the minimum number of components needed to understand the management structure of today's cellular networks. This figure contains the following components of a simple cellular network:

Base Station (BS). This is the radio transmitter and receiver communicating to the end user's wireless device. These stations are arranged in cells to maximize the spatial reuse.

Base Station Controller. This controls many Base Stations. It is responsible for managing the handoff of calls from one cell to another as the end user moves between cells.

Mobile Switching Center (MSC). This is the switch that controls many Base Station Controllers. It is similar to a standard switch in the PSTN, but it has additional functionality to handle mobility of the end user. The MSC works with the Home Location Register (HLR) to track the mobile user.

Home Location Register (HLR). The function of the HLR is to keep track of where the mobile user is. It helps manage user mobility by communicating with the MSC to always know what MSC a user is currently connected to.

One reason this structure is centralized is because the network tracks what the users are doing. The network knows what users are doing, no matter where they are. With a cellular network, servers within the network must know the location of the end device even when it is the end device initiating network communication. In effect, cellular networks impose a centralized structure on the distributed structure of the Internet access they provide.

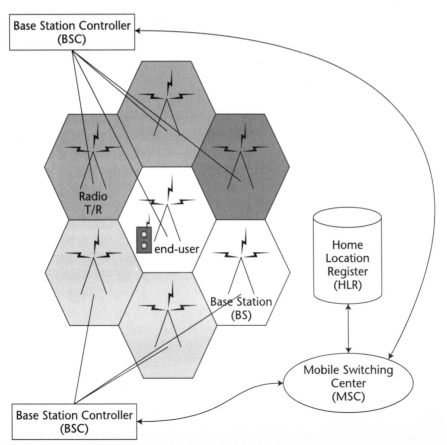

Figure 11.2   Simple cellular network infrastructure.

There are many advantages to this centralized management structure. The spectrum, equipment, and network resources are used efficiently. Users can have a one-stop shop where all billing is consolidated. Well-engineered cellular wireless networks are an efficient way to utilize scarce resources. As a result, there are substantial economic incentives for profit-motivated firms to undertake on the operation of such services. This efficiency, though, must be weighed against the disincentives to foster innovation.

802.11: Technology

As previously mentioned, networks based on the IEEE 802.11b standard, also know as Wi-Fi, are popping up everywhere — in homes, offices, hotels, conferences, airports, and other hot spots. Individuals, loosely organized groups, and enterprises are building such networks because it is easy, it works well, and it is quite inexpensive. Anybody with an Ethernet connection to the Internet (including DSL/cable modems) can plug in an 802.11 access point to this Ethernet connection and have broadband wireless Internet connectivity. Its really that simple; I have such a wireless network in my office and my house.

In a few locations like Cambridge, Massachusetts, that have a high density of wireless access points, ad-hoc community Wi-Fi networks have evolved because individuals keep their access points open. While there is no guarantee of quality, some would venture that the proliferation of Wi-Fi is close to tipping toward widespread adoption. A viable economic model for commercial deployment, however, has yet to emerge.

The technologies of 802.11 are a group of standards specified by the IEEE and classified as a low-power, license-free spread-spectrum wireless communication system [5][6]. License-free spectrum means anybody can use it, but this spectrum must also be shared with other devices (for example, microwave ovens and X10 wireless cameras). Originally, the 802.11b standard was introduced with a 2 Mbps data rate, later being increased to 11 Mbps. It uses both frequency hopping spread spectrum (FHSS) and direct sequence spread spectrum (DSSS). Table 11.2 illustrates the 802.11 standards [3][5].

802.11 technologies are designed for hot spots, but they may turn out to work well in some wide-area applications. By increasing power or with external antennas, the range of 802.11 can extend for miles [6], which means Wi-Fi is a viable last-mile technology. Covering neighborhoods and business parks with 802.11 access points might be a cost-effective method for providing broadband wireless services in homes, offices, business parks, and public places.

Table 11.2 Characteristics of 802.11 Technologies

EMPTY	SPECTRUM	MODULATION TECHNIQUE	BANDWIDTH	DISTANCE
a	5 GHz	OFDM	54 Mbps	60 (feet)
b	2.4 GHz	DSSS (WiFi)	11 Mbps	300 (feet)
g	2.4 GHz (b)	CCK-OFDM	11, 54, (a) and (b) compatible	300 (feet)
	5 GHz (a)			60 (feet)

802.11: Management Structure

802.11 is an adaptable technology allowing for flexible management structures. At one extreme, it allows a completely distributed structure where individual users at home and in the office install and manage their individual access points. Such networks can provide temporary IP addresses with a DNS/NAT where nothing within the Internet knows about this connection, which makes tracking its usage impossible. At the other extreme are large service providers that install and manage Wi-Fi networks for many different individuals or organizations, such as for wireless access in airports. At the middle ground are organizations that install and manage their own Wi-Fi infrastructure for internal use, such as college campuses and enterprise facilities. Wi-Fi networks have the flexibility to be managed either centrally or in a distributed manner.

These Wi-Fi networks are simple to install. When individuals buy, install, and manage their own wireless access points, the overall hardware/software and management costs can become substantial. In this completely distributed architecture depicted in Figure 11.3, both Mark and Nalin have 802.11 access points in their homes; in Mark's case, his coverage overlaps a neighbor's access point. Mark and Nalin also have access points in their offices. These office access points are within the Boston University, School of Management (SMG) network, but are not managed by the IT department of SMG. In both cases, these access points are managed by the individuals who are the primary users, leading to inefficient use of resources.[2]

---

[2] Interestingly, it is hard for service providers to tell if a user with wired access has installed his or her own wireless access point.

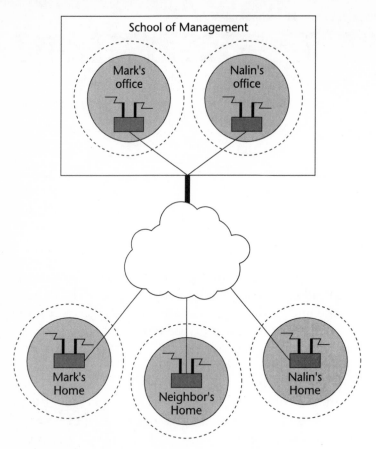

Figure 11.3   Individually managed Wi-Fi access points.

802.11 networks are amenable to a more centralized management structure. I installed both my wireless points (one at home, one in the office)
because doing so is the only way to get the wireless service I want. I would
prefer that the IT department at SMG provide wireless service within my
work environment and that my cable company provide wireless service in
my house. Imagine the convenience of cable or DSL modems that came
with built-in 802.11 access points or of the cable/phone company providing the last mile into our homes with a wireless access point located at the
curb, as depicted in Figure 11.4. With this structure a service provider is
providing the last mile into our homes with an access point located at the
curb. I would gladly pay the service provider for both basic Internet service
and the wireless connectivity. This more centralized management structure
is more efficient and can potentially provide connectivity for more users. In
fact, I believe that firms with network management expertise, such as IBM,
will find a profitable opportunity to provide and manage such networks
for enterprises.

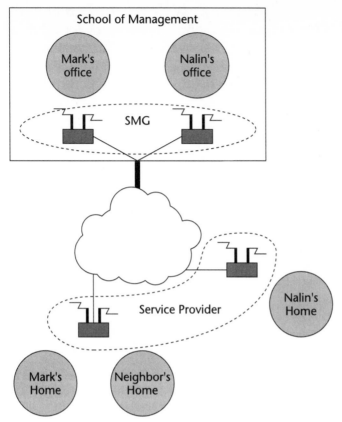

Figure 11.4   Campus Wi-Fi hot spots.

It is conceivable that a single large service provider will provide coverage to a large metropolitan area such as Boston (Figure 11.5). This is the most efficient way to manage 802.11. Clearly, it's the least expensive structure that can be used to provide service to a large number of users in densely populated centers; it's also the least flexible. With only a single service provider, innovation is slow, and users have no power to bargain. Again, there are trade-offs: the efficiency of a very centralized architecture compared to the loss of flexibility.

Currently there are several models of providing wireless Wi-Fi services with a more central management structure than the ad-hoc approach of setting up your own access point while being oblivious to what others are doing. Community groups such as Bay Area Wireless in San Francisco and NoCatNet in Sonoma, California, are organizing to provide free access. Joltage and Boingo's business models are more centralized — they allow small organizations or individuals to become access points to members of

Joltage's or Boingo's community. Members pay Joltage $1.99 (as of April 2002) per hour or $24.99 for 60 hours per month. In return for providing a wireless access point as part of its network, Joltage pays the access point's owner 50 percent of the revenue it generates — a cool model. Joltage manages the billing and customer management; the access point service provider manages only the wireless and Internet service. Boingo's model is building 802.11 access points in airports, hotels, and cafes. Boingo users pay up to $74.95 per month for unlimited access, a good deal if you happen to live at hotels that provide service. The evidence of Wi-Fi's flexibility in management structure are the many management models that have emerged — from the most distributed ad-hoc un-coordinated structure to centralized service providers such as Boingo.

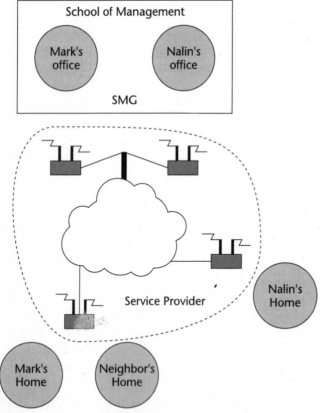

Figure 11.5 Metropolitan area Wi-Fi networks.

As explained in the previous text, 802.11 is a versatile technology allowing a continuum of management structure, from extremely distributed to highly centralized. The more distributed the structure of this management is, the more conducive it is to experimentation, but at the cost of efficiency. The more centralized the management structure, the more efficiently resources are used, but at a cost of no user control or ability to experiment. This illustrates why 802.11 is becoming so popular — its flexibility in choice of management structure meets the needs of many users, from cowboy technology geeks wanting freedom to experiment, to conservative IT managers in the largest organizations wishing to constrain what users can do.

Application to 3G and 802.11

This section applies the aforementioned theory to the wireless data Internet access technologies — third-generation cellular and 802.11 LANs (Wi-Fi) — to predict how the future might unfold. I demonstrate how value can be captured when third-generation cellular service providers gain minutes of airtime or packets of data transfered from services provided by independent service providers. I illustrate how service providers can capture value from innovations in the 802.11 world by learning what services users want in the 802.11 environment. These services can migrate to 3G cellular networks from other types of networks where they are discovered and proven successful. It is expected that new, innovative services will come from many places, not just from the service providers of cellular networks. Large centralized wireless service providers can gain additional value by managing 802.11 network infrastructures for small distributed wireless service providers and by centralizing the billing of these distributed 802.11 services. The value of users having choice and the economics of externalities imply that the coexistence of 3G and Wi-Fi makes sense in the future.

The previously mentioned theory illustrates the potential value of experimentation by the distributed management structure possible with 802.11 technology. It also demonstrates how centrally managed cellular wireless networks can capture this value of innovation by learning about what services work well in other wireless environments and by using the business and technical advantages of central management to efficiently provide services that have been proven successful. This theory explains how innovation in experiment-friendly environments is required to find the services users want and how centralized management is better able to efficiently use the limited resources inherent with wireless networks. By coexisting and building on each other's strengths and weaknesses, 802.11 and next-generation cellular networks will combine to create great value because experimentation in distributed networks will allow users to get services that meet their needs and the efficiencies of centralized management will allow these services to be priced so many users can afford them.

From this theory, if market uncertainty changes in one direction or cycles between high and low, as expected, then a flexible network infrastructure is required. When market uncertainty is high, then it is expected that the distributed versions of 802.11 will work well. Uncertainty is expected to be high for many years to come with wireless applications. It is also expected that certain services will be successful and will be prime candidates for a centralized management structure after they have become successful as services managed with a distributed management structure. Because market uncertainty will be very dynamic, increasing for some services, decreasing for other applications, it makes sense that both the distributed and the centralized management structure will coexist.

Effects of Experimentation

For now, nobody knows what the potential is for wireless services or who will create the best of them. We do know that Web browsing on micro screens is not the answer, but what is? Lots of experimentation is needed to figure this out. Users, graduate students, vendors, and service providers need to think of new ideas and determine whether users like them. This means applications will likely be developed on traditionally wired networks or maybe 802.11 wireless networks because this is the environment in which most Internet innovations have occurred, and will continue to occur, because of the inexpensive and easy-to-use equipment now available.

Ideas for new services will come from everybody — the more ideas for services, the better for users. Wireless service providers such as telcos do not have the breadth of knowledge to innovate the best services in all areas. While telcos are good at creating successful traditional voice services, they are not good at innovating new services in the financial areas or other areas outside their core competencies, as illustrated in Table 11.3.

The rows correspond to the type of services, such as location tracking, where people or equipment are tracked, or radio and video services such as IP radio or video conferencing. Each column represents an industry type, such as the transportation industry or the telephone companies. Each entry is how well we expect experiments from one particular industry to fair with services of a particular type. For example, we expect companies in the financial world to do well when experimenting with financial applications, but poorly for all other types of services. Allowing those with the most experience in their area to experiment within these areas creates the most value.

Figure 11.6 illustrates the effect of Table 11.3 on how the value of experiments will be distributed. This figure shows normal distribution curves for different means and variances. Firms with average information about a particular industry are expected to have average performance in the context of experimentation with new services. This is indicated by the centered distribution with a mean of zero and variance of one. As a firm becomes expert in a particular service area, the expected value of its distribution shifts to the right and the variance decreases because the experiments should be more successful and focused, thus meeting the market better because of the firm's added expertise. As organizations experiment in areas they know nothing about, the mean is expected to shift to the left, reflecting the inexperience of the firm in this new area, but the variance should increase, indicating that if they are extremely lucky, they might find a great application. This figure illustrates how when firms become experts in a particular area, their experimentation is more focused and successful.

Table 11.3   How Expertise Changes the Value of Experimentation

SERVICE TYPE/ INDUSTRY TYPE	TRANSPOR- TATION	FINANCIAL	TELCO	ENTER- TAINMENT
Location tracking	High	Low	Medium	Low
Radio/video	Low	Low	Medium	High
Traditional voice	Low	Low	High	Low
Financial trading	Low	High	Low	Low

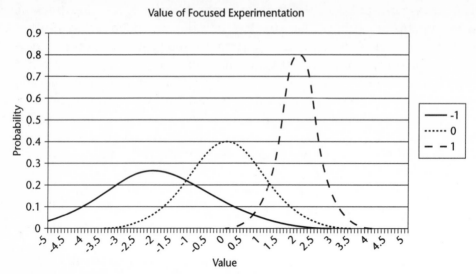

Figure 11.6 Shifting the mean and the variance.

Figure 11.7 is a surface illustrating this landscape, representing the value of firms experimenting with new services. The X-axis corresponds to the X-axis, and the Z-axis represents the Y-axis of Figure 11.6. The Y-axis gauges the knowledge of the firm in the particular area. This ranges from +1 for firms that are experts to –1 for clueless organizations; zero indicates organizations with average information. As y increases from 0 to +1, the mean increases and the variance decreases. Accordingly, as y decreases from 0 to –1, the mean decreases and the variance grows. This figure demonstrates that there are two ways to generate great applications: (1) lots of experiments from nonexperts or (2) a few experiments from industry experts. The most benefit comes from firms experimenting in areas they know best — the region in the back right corner. This high peak represents the optimal region in which to experiment. The most value is created when services are allowed to come from places that understand the application because they leverage the previous experience in a market.

Building services that are unaware of the underlying network is important to capturing the most value in the coexistence of 802.11, next-generation cellular networks, and the rest of the Internet. Applications unaware of the underlying network allow a seamless migration from one network infrastructure to another. The most value is created when services developed on one particular network architecture can migrate to another network infrastructure without changing the application. This migration of services allows the value of experimentation from distributed management to be captured by networks with a centralized management structure. These applications that don't care about the underlying network structure create the most value because these services can seamlessly migrate between the different network architectures.

802.11 — Advantages and Disadvantages

The value of experimentation with 802.11 is tremendous. It is so easy to do and so inexpensive. Students in many schools do wireless projects. The point is that there is so much experimentation going on and the market uncertainty is so high that one of these days (soon, I hope), someone will discover a killer application, or at least enough niche applications to keep service providers alive. This combination of high market uncertainty and easy experimentation is very valuable in the context of meeting uncertain user needs.

The 802.11 set of standards is very flexible in what management style the standards allow, as discussed in this chapter. This trait of allowing choice in management structure is valuable because it allows different users to get what they want. Some users will choose to manage the service themselves; others will choose to buy a bundle of equipment and services. Users, vendors, and service providers will be able to experiment, thereby giving users many choices. As the number of choices increases, or as the market uncertainty grows, this value increases. Users are best off when they have choices of several flavors of wireless services — from flexible distributed to efficient centralized structure. Having a management structure that can fall along a continuum between a distributed and a centralized management structure works well when market uncertainty is dynamically changing, and this is what 802.11 allows.

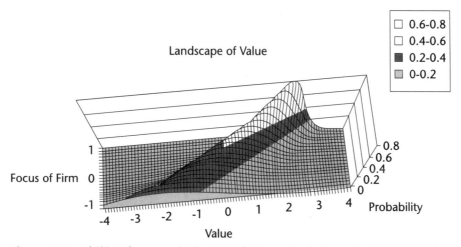

Figure 11.7   Shifting the mean/variance surface.

The distributed management structure that 802.11 technology allows is very flexible because users can experiment with technology they own and manage. Owning and managing the equipment are conducive to experimentation because no central authority controls the users, which means they don't need permission to experiment. 802.11 networks designed with distributed architecture give users power because they can do what they want, when they want.

For users with business models depending on having the most advanced technology, control over when to upgrade the technology is vitally important. With distributed management, when new technology is introduced the user can decide when to upgrade to the better technology without depending on the service provider or IT department. When 802.11 technology started to support the 11 Mbps data rate, if you didn't manage the access point and end device, then you didn't get to decide when to move up to the faster technology. It's nice not to worry about your wireless service, but for the most demanding users this may be a frustrating experience.

When service providers or the organization's IT department manage the wireless access points, each user has less to worry about but also less control over the wireless service. This more centralized setup gives users less ability to experiment because they need permission from the network manager. Furthermore, the IT department or the service provider decides when to upgrade to a newer technology, which may not meet the needs of the most demanding users.

Overall, having one access point in every home and office is, in total, the most expensive and time-consuming way to provide wireless service. This means I do my own installation, configuration, and maintenance of the access point. While each individual access point is easy to install and manage, there are so many of them that the aggregate effort is great. This approach is inefficient in its use of equipment — at my house my access point never has more than my three family members; in my office, I am the only user. While not efficient, this structure works well, which is why it is creating a large user base.

The preceding analysis, which illustrates the value of 802.11's flexible infrastructure allowing a continuum from a distributed to centralized management structure, is similar to the argument I made in Chapter 10 about the value of SIP because its flexibility allows services to be built with either an end-2-end structure or a centralized architecture. SIP allows end-2-end users to experiment, similar to how users can set up an 802.11 network infrastructure to experiment with. SIP also allows services to have a centralized structure via SIP proxy severs, similar to how 802.11 allows a large service provider to supply this technology with efficient central management.

SIP allows both environments to coexist, similar to how many structures will coexist with 802.11 wireless (and cellular) services. While SIP and 802.11 are completely different technologies, they have amazing similarities in the flexibility they offer regarding management structure because they both allow a continuum of architectures from distributed to centralized.

Next-Generation Cellular Networks

Cellular network infrastructure, while not as flexible as 802.11 technology, does have some business and technical advantages. Cellular technology has a very different goal from 802.11 technology — 3G aims to provide wireless broadband services in a ubiquitous manner with a centralized management architecture, which it does very well. This is different from the current uses of 802.11, which is focused on coverage of hot spots. In its current form of deployment, 802.11 will not make sense in sparsely populated areas; the economics just don't work out. One of 3G's main business and technology advantages is its ability to cover large areas efficiently; with its centralized management structure it can efficiently provide these services to many users at affordable prices.

The intelligent and centralized structure of cellular networks allows service providers to know what users are doing. This has such advantages as better security and better Quality of Service (QoS). The centralized control inherent with current (and next-generation) cellular networks bodes well for good security and grantees of service because the network is aware of what all users are doing. The dreaded Denial of Service (DoS) attacks that have become popular are much more difficult to launch and sustain when an intelligent network knows what its users are doing. This information — what the network knows about its users and what they are doing — also makes QoS less complex. One advantage of the centralized network infrastructure is that when you watch users, you can also protect them and guarantee what they can do.

Most industry experts agree that large centralized cellular service providers are good at billing customers. One strong argument for large centralized service providers is the convenience of one-stop shopping. These cellular service providers are perfectly set up to become centralized billing agents because of their centralized control and smart network design. Many of the wireless providers have roots in the traditional phone networks, which means they are good at keeping track of what users are doing and billing them appropriately. Many users value the ability to roam anyplace in the country (and the world with tri-band phones) and use many different services from many different service providers, while receiving a single bill. Cellular providers are good at reaching agreements with other service providers, giving users the convenience of a single bill containing fees for many different services. Large cellular service providers are likely to succeed with this business model of consolidated billing information from many services because it is what they have been doing for many years.

Networks with a centralized infrastructure, such as cellular, are very efficient at equipment usage. Deciding how to deploy wireless access points covering larger areas is a complex decision requiring advanced engineering techniques [3]. There are many factors to consider, such as the local geography, man-made obstructions, and interference from other sources. Efficiently deploying equipment requires centralized planning, which is inherent with cellular networks. Individual users in neighborhoods and users setting up rough access points in office environments will find it impossible to coordinate their decisions to implement the most efficient network infrastructure. Networks with a central management structure will always use the limited resources to their best advantage because of the ability to centrally plan their networks.

Installing equipment is only part of the cost of a wireless infrastructure because it must be managed, which turns out to be expensive. The centralized management of cellular networks makes this easier because workers can be trained and equipment can be standardized, which is unlike the heterogeneous equipment used with the 802.11 infrastructure where each user decides which vendor's equipment to use. Standard equipment and widespread deployment simplified these management aspects. When managing equipment, the centralized structure of 3G has value.

Spatial Reuse Efficiency

One important difference between the 802.11 and 3G technologies is ownership of spectrum. The spectrum with 802.11 is public and must be shared with other users, while the service provider owns the spectrum with cellular technology. Ownership of spectrum creates value by locking in customers because nobody else can use this spectrum. But, is this lock-in worth the cost of the spectrum? How great a handicap is public spectrum? These important questions remain unanswered.

Spectrum ownership decreases network implementation uncertainty because ownership means control over its use. This uncertainty has two forms. First, there is uncertainty about your neighbor's spectrum use because it might interfere with yours, and because you both have the same rights to the spectrum, you can't do much. There are many other uses for this public spectrum, such as X.10 cameras and microwave ovens, with equal rights as 802.11 to use this spectrum. The second type of uncertainty is what type of government regulations might affect your network build-out. The benefits of spectrum ownership are clear, but the value of this ownership is not.

Wireless transmission is hard because many factors interfere with the propagation of signals. The centralized structure of cellular networks fits well with the large-scale planning required to efficiently place the wireless Base Stations. Spectrum is very expensive, which implies that service providers must maximize the number of users in a region. This is possible only with careful positioning of the Base Stations to maximize spatial reuse. Large centralized cellular providers are in an ideal position to design and implement efficient wireless structure because they have the rights to spectrum and typically have some leeway as to where Base Stations are located.

There is a trade-off with 802.11 and cellular services in the context of spectrum versus the management efficiency of the centralized structure. Spectrum can be expensive, but if you own it, you can apply advanced engineering techniques to use it more efficiently (most bandwidth for the most users). Planning how to reuse the limited spectrum most efficiently over large areas is key to affordable cellular services because spectrum is such a limited resource. Spectrum can't be manufactured, so business models demand the most efficient use of this most critical scarce resource. Is this ability to centrally plan and manage your spectrum worth the cost of the spectrum? Even though 802.11 spectrum is also limited, it is free. Are the inefficiencies of managing 802.11 spectrum overcome because this spectrum is free? These are complex questions that will be addressed in future research. This chapter points out the complex relationship between cost of spectrum and management cost of the spectrum — what is the value of free, hard-to-manage spectrum compared to expensive, easy-to-manage spectrum?

The preceding text illustrates the complex trade-offs between 802.11 and 3G. The case for coexistence is strong. Compatibility between 802.11 and 3G is essential to capturing the most value from network services in general and wireless network services in particular. Figure 11.8 is one view of the future world. It demonstrates the ubiquitous nature of current and next-generation cellular technology. The entire region is blanketed with cellular service. This works everyplace, in buildings, subways, and outside the city. It is not the highest speed, but the ubiquitous coverage makes it a valuable technology. There are other spots where the economics determine that it makes sense to build a higher-speed, and more expensive 802.11 infrastructure. Locations such as airports, offices, and even homes are examples where this approach works well. Even business parks and dense neighborhoods might be good candidates for 802.11 wireless coverage. There are many places where wired connectivity makes the most sense. My desktop systems at home and in the office fit this category nicely. They never move, and I want the highest bandwidth and security, which the standard Ethernet-switched LAN networks provide. This figure demonstrates that there is room for all three technologies, and all three will coexist and thrive — each service creating value for the other services because of the coexistence of these technologies.

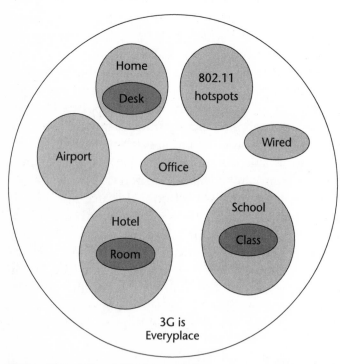

Figure 11.8   3G and 802.11 coexisting.

Conclusion

This chapter addressed three main economic factors important to the argument for co-existence of $N^{th}G$ cellular and Wi-Fi technology: tipping, externalities, and increased user market caused by complementary technologies. Tipping occurs when users jump on the bandwagon with everybody else, an act that accelerates how new technologies are adopted. The economics of externalities illustrate how value increases to each user as the number of users grows. When this is coupled with the evidence that the market is tipping, the increased number of users caused by interoperable technologies, the value of $N^{th}G$'s friendly coexistence with 802.11 is exacerbated. The economics of viewing Wi-Fi and $N^{th}G$ cellular as complements, not substitutes, are strong because both users and service providers win.

To gain the most from network effects, the user market must be as large as possible. By encouraging interoperability between wireless technologies, the total number of users will be larger. More total users equates to more users for each technology, as users will tend to use both of them, picking the technology that fits the particular situation, as long as the business model supports this interoperability. The effects of coexistence are illustrated in Figure 11.9 — it is helpful to both technologies. The X-axis is the number of users; the Y-axis is the value to each user of the service. N_1 and N_2 are the number of users of 3G and Wi-Fi wireless service treated as substitutes. The bottom curve is the value of 3G, and the top curve is the value of WiFi; as expected, as the number of users grows, so does the value to each user. The star is the value of having N_1 users. More users implies more value, and when cellular service providers can tap into the Wi-Fi market, the number of users increases to $N_1 + N_2$, and the value increases to $3G_C$, the dot. This figure illustrates this because $3G_s < 3G_C$. The same effect is seen on the Wi-Fi value curve — $WiFi_s < WiFi_C$. More users equates to more value per user. Thus, the value of complementary technologies is: $(3G_C + WiFi_C) - (3G_s + WiFi_s)$. In the case of these technologies, cooperation has more value than competition because of network effects.

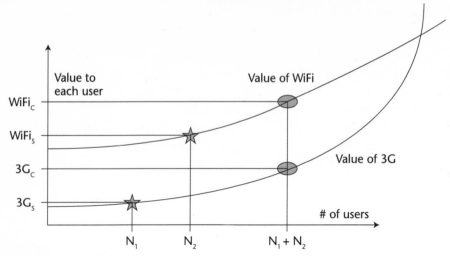

Figure 11.9 Value N^{th}G and Wi-Fi as complementary services.

To maximize value, the larger user base needs many choices. Service providers must come from every nook and cranny if the best services are to be found. As discussed before and illustrated in Figures 11.6 and 11.7, different groups will have different expertise. Large telco-like service providers need to capitalize on the massive innovation that occurs when niche players experiment in their area of expertise by looking everywhere for new services. The most value comes from giving users the most choices because some of these choices might be superior matches to the uncertain market.

The current business model of cellular service providers seems the opposite of what is needed to capture value from the next-generation cellular data services provided by smaller, niche providers specializing in certain markets. Today's wireless service providers market plans that encourage most users to buy more minutes than they use. This model won't transfer well to cellular data services because if these wireless service providers gain additional minutes only from independent service providers, they lose; they get no additional revenue as users' usage increases closer to their allowable allocation. As users begin to use up their minutes with services not directly related to the wireless provider, it costs the wireless provider more, but generates no additional income. This current business model won't work because services from providers unaffiliated with the wireless provider will increase costs to the wireless provider without generating revenue to offset the increased cost. This is the opposite of what is needed — a way for unaffiliated service providers, affiliated service providers, and the wireless service providers to all share in the success of wireless services.

I-mode, a next-generation cellular data service started in Japan, is succeeding with its business model. I-mode services are of three main types: those offered by the wireless service provider (DoCoMo), services offered by other vendors but affiliated with I-mode because it manages the billing, and services that have no association with DoCoMo — they only provide extra data transfer charges to them. Models like I-mode, in which users pay as they go, may provide the economic incentives needed to convince the wireless providers to open up their networks, similar to what DoCoMo is successfully doing in Japan. By adapting their current business model to promote independent service providers, these wireless service providers can capture the most value.

This chapter illustrates the link between market uncertainty and management structure with wireless services that previous work [1][7][8] predicts, which will illustrate why 802.11 and next-generation cellular technologies will coexist — each adding value to the other. For now, the needs of users seem mysterious to vendors, service providers, and even users who are unable to articulate what they want. The flexibility of the 802.11 technology allowing distributed structure has great value when market uncertainty is high — it is so easy to build infrastructure and experiment with new services. 802.11 technology can also be implemented with a more centralized management structure, taking advantage of business and technical advantages. 3G technology has benefits because it has attributes that 802.11 cannot currently compete with — spectrum ownership and ubiquitous coverage. 802.11 and 3G network infrastructure will coexist, each gaining value from the other.

The argument utilized an options framework illustrating the value of 802.11 technologies because it allows users to experiment and illustrates how next-generation cellular networks can capitalize on the innovation occurring in 802.11 networks. I previously discussed the value of how services migrate from one network infrastructure to another. This is based on the real options approach, and it is something to investigate in future research. The main point in this book is that giving users as many choices as possible when market uncertainty is high creates the most value. These choices need to include both what the applications are and how the user accesses them. Higher market uncertainty results in a greater value in giving users many choices. This real options framework is one way to quantify what seems to be happening with the adoption of 802.11 technology from the ground up.

This chapter argues for the coexistence of many technologies to give users broadband Internet access. It illustrates the value of having different network infrastructures to meet different needs — from wired, to hot spots, to ubiquitous coverage. It explains how coexistence will create the greatest overall value — the biggest pie. My model includes large centralized cellular providers as providers of both cellular and 802.11 connectivity. They are in a unique position to centralize hot spot broadband access and effectively bill users for it. They can also enter into service agreements with smaller distributed 802.11 service providers and provide customer management and billing services. Everybody has cell phones and receives bills; these bills seem an obvious place for wireless Internet access fees. I argue that applications developed with one infrastructure will migrate to others and create more value than a single architecture allows because coexistence gives providers of both types of service a piece of pie that is bigger than the entire pie if both technologies compete with instead of complement each other.

The next chapter is about Web applications and services. Some Web-based applications, such as auctions or information dissemination, are very successful; some, such as buying furniture online, are not. Web applications have changed everything for both users and companies over the last 10 years. Similar in some ways to Web applications are Web services based on the Simple Object Access Protocol (SOAP) and Extensible Markup Language (XML). The name "Web service" seems to have stuck, but these network services based on SOAP and XML standards have little to do with the Web. Both Web services and applications display similar attributes in the context of choice in management structure. Web applications and services allow the easy experimentation of distributed networks, and they can also capture the business and technical advantages of centralized management. Web applications and services fit well with the theory from Part One.

CHAPTER 12

Web Applications and Services

This chapter discusses Web-based applications and Web-based services. Web-based applications refer to services that users (mostly people, but maybe bots, agents, and screen-scraper applications) access via a Web server. While automation is possible, most Web sites are designed for people, not computer programs, performing a task. Examples of this type of Web-based application include email, as described in Chapter 8, eBay, the popular Internet auction site, and MapQuest, the successful direction service. Some types of Web services (as they are known today) are similar to Web applications because of the protocols they use, but Web services are more focused toward distributed application software exchanging data based on Simple Object Access Protocol (SOAP), Extensible Markup Language (XML), and other standard protocols. In addition, Web services assume that they are exchanging data with a software client that conforms to the Web service's defined interface. Both Web applications and Web services are similar in the range of management structure they allow users and service providers, and they are both very flexible because of their architecture and the extensibility of the standards on which they are based.

Many people confuse Web-based applications and services with network-based services. All Web-based applications and services are network-based services, but not all network-based services are Web-based. Two examples of this are the traditional voice services provided with

Centrex or PBXs, or email. Email may or may not be Web-based; it depends on the email client and server. Techies reading email with Pine or Emacs, or users of Eudora, or office environments such as Lotus Notes are not Web-based. Web-based email, though, is becoming more popular (as discussed in Chapter 8) and is available as both distributed and centrally managed services. Generally, Web-based applications and services have the flexibility to benefit from attributes of a distributed and a centralized management structure. There are many flavors of Web-based services, and for now they fit into my analysis of general network-based services.

The name "Web services" confuses many people, including me, because while some Web services are similar to Web applications, other Web services have little in common with the Web. A Web service using HTTP to exchange SOAP messages is similar to a Web application because they are using the same application protocol to exchange information. Web services, however, do not demand the use of HTTP and are perfectly happy with other application transport protocols such as SMTP and FTP. A Web service using FTP to transport SOAP messages has nothing to do with Web applications. To avoid confusion in this chapter, I rename Web services to what I believe is a more correct name: soap services. It is thusly named because of the SOAP envelope that Web services utilize to transport messages between users. The name "soap services" fits the wide range of Web services.

While different, Web applications and soap services have some similarities. Both of these technologies give users the ability to experiment. Many tools for Web applications and soap services development come standard in Unix environments or are open source. Both of these technologies are based on standards that are open and freely available on the Web. This reliance on open standards that are freely available promotes experimentation and innovation by users with both Web applications and soap services.

Another important attribute that Web applications share with soap services is flexibility in the management structure they allow. Because of the end-2-end nature of both Web applications and soap services, it is easy to experiment with these technologies. Without asking the network service provider for permission, any two users on the Internet can implement a Web application or soap service without any changes to the network infrastructure. These two users are the only ones that know about this new experimental service, which means that nobody else on the Internet is affected by this experimentation. Once used by a large group, though, these same Web applications and soap services can display attributes of centralized management. When it is in widespread use, changes to the application or service do affect a big user group, which makes it hard to experiment with. This flexibility of Web applications and soap services to begin their life with a distributed management structure and then morph into more efficient centralized management is of great value in dynamic markets, as the theories from Part One discuss.

Web Based Applications

The Web has proven to be a flexible platform for many tasks. It seems everything is becoming Web-based — sometimes with wonderful results. Banking at home on the Web or having Web-based access to information about my students saves time — putting valuable data and useful services at my fingertips. Some Web-based applications are great for users and service providers. eBay is one example; users love it because of its ability to put buyers and sellers together. It also has a good business model because the service provider (eBay) makes money. Other Web-based applications are well suited for users but have yet to develop business models that support their continued existence. Examples of these are search engines and map/direction applications that are popular with users but may not generate enough revenue to sustain themselves. Still other applications have proven to be of little value to service providers, users, and investors — as illustrated by the "dot-com" crash. Thus far, the Web and Web-based applications have evolved far beyond anything their architects imagined, creating and destroying vast fortunes, transforming business processes, and changing society.

Sometimes the idea of a Web-based application is good, but the implementation and/or the user interface is not. How many times have you given up hope of an online transaction and instead called the organization to complete your task? The interface is really bad when you can't even figure out how to call them. In this context, Web-based applications are similar to other application software — ultimately the user needs to accomplish his or her tasks. Web-based applications that are not easy to use fail despite the value of the underlying application.

Web-based applications allow flexibility in management structure. As discussed in Chapter 8, centralized Web email exists (Hotmail), as do more distributed models provided by ISPs running the POP protocol with a Web interface. In general, Web-based applications are able to take advantage of the attributes of centralized and distributed management. This is true in several ways: user domain, management of data, and the ability of end users to experiment. The email example explains how Web-based applications such as Hotmail have a distributed group of users that cross many user domains, while other Web-based email applications have a more limited domain (see Chapter 8 for details). The email example also illustrates how Web-based email services may have both distributed and centralized management in the context of data management: Hotmail users manage messages on the mail server (a centralized structure), but ISP POP-based Web email systems download messages to the user's machine (a distributed style of data management because the user manages the messages). Web-based applications allow user experimentation because of the end-2-end nature of Web protocols. Once large and successful, these Web-based systems become harder to change due to their broad base of users; this is an attribute of more centralized architecture. Thus, Web-based applications have the strengths of both the distributed and the centralized management structures.

Web-based applications can be developed in a very distributed manner because setting up a Web server and experimenting with Web pages and applications are trivial. It's possible to keep the URL of the Web-application completely isolated from everybody else in the world. It's also possible to limit a Web application to a narrow group of users, such as the faculty link at Boston University, which allows faculty access to student information, making this a distributed architecture with a well-defined group of users. Changes to this system affect a known group of users. This is unlike an application such as MapQuest that provides maps and driving directions to a general audience, which is a more centralized structure and therefore harder to change. An environment that allows easy experimentation is exactly what is needed to find Web-based applications that meet users' needs due to the high uncertainty in this new area.

Soap Services

The questions companies ask, such as, "How should I integrate my many different IT systems" or "How can I maximize my return on investment (ROI) when building new IT applications?", seem to have the same answer — soap services. In theory, soap services allow interoperability between heterogeneous systems, but it's too early to determine if this lofty goal will reach fruition because of the high level of market uncertainty. Most vendors (Microsoft, Sun, IBM) agree about the high-level aspects of soap services — they are based on open standards such as XML and SOAP. The architecture of soap services promotes modularity and easy experimentation because they can be designed with end-2-end structure, which, as discussed in Chapter 3, promotes user experimentation. Currently, the most debated details focus on the server, rather than the client, and include what platform is best to build soap services on and what language is best for developing soap services. Fortunately for users of soap services, the value of these services depends on the architecture of the services, not the platform on which they are implemented. While the future of soap services is unclear, it is clear that they facilitate integrating vastly different information systems.

Everybody agrees on the need for infrastructure that permits easy implementation of and experimentation with new applications. There is no agreement, though, on how to reach this goal. Microsoft believes in its .Net strategy, but Sun, IBM, and others don't share this vision. For my argument, it doesn't matter what the specific technology will be, as long as it allows users to develop new applications. For now, everybody agrees on the big picture about how to exchange information using standard protocols such as SOAP and XML between any two pieces of software running on any two computers, as long as they are connected on a network and follow the standards.

Soap Service Background

Soap services provide a standard way for heterogeneous systems to exchange information. As discussed in the text that follows, soap services are not a new idea, but rather they are the continued evolution of many older ideas. What is new about soap services is that they are based on open standards that are accepted by everybody. These standards include the message structure (SOAP), how data is encoded (XML), the API detailing the interface that explains how to use the service (WSDL), and how to register a service so others can discover its existence (UDDI). At the highest level, a soap service is a resource on some computer invoked by sending it a message following the SOAP protocol. This SOAP message contains the information needed to perform the soap service. For example, Figure 12.1 illustrates a simple soap service to report the temperature for a city. This weather-service is listed in the yellow pages of soap services, along with directions for its use. The client can look up the soap service and call it to find the temperature in Cambridge, Massachusetts. Both the request and the response are sent in a SOAP message. This information is encoded in a standard way: XML. Because all soap services agree to use SOAP and XML, they allow any client access to these soap services as long as the client follows the standards. This is very similar to how a Web browser gives users access to Web applications. Soap services are not a new idea, but rather, the next generation of Remote Procedure Calls (RPC) [1]. Soap services promise the Holy Grail to IT professionals: a standard way to link any two systems so that they can exchange information.

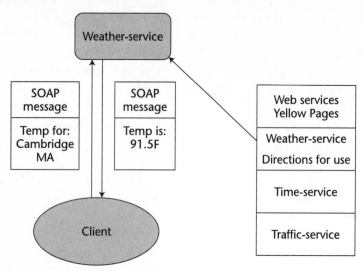

Figure 12.1   Basic weather SOAP service.

What Is a Soap Service?

The definition of a soap service is vague. Different vendors use different languages, but most agree about the basic structure and definition of soap services. While these definitions from a few major players are different, they have a common thread — soap services are network-based services implemented with open standards that allow different systems to communicate. A broad definition of soap services is this: *a piece of executable code identified by its URI that is coupled to its client via a SOAP envelope containing XML encoded data.* In many cases, this code performs a piece of business logic. Any client on the Internet can access this piece of business logic as long as the client implements the open standards SOAP, XML, and other Internet protocols. Soap services provide a way to expose fine-grained business logic within an organization, as well as externally to business affiliates. One attribute that all vendors agree on is a loose coupling between client and service with open standards; this gives clients, on any operating system and any hardware, the ability to consume a generic soap service. The IBM Web site describes this server/client independence [2]:

> *A Web service is an interface that describes a collection of operations that are network accessible through standard XML messaging. . . . The nature of the interface hides the implementation details of the service so that it can be used independently of the hardware or software platform on which it is implemented*

and independently of the programming language in which it is written. This allows and encourages Web services-based applications to be loosely coupled, component-oriented, cross-technology implementations. Web services can be used alone or in conjunction with other Web services to carry out a complex aggregation or a business transaction.

Soap Services in Context

The simple idea behind distributed computing environments is as old as networking: utilizing a computer at one location from another. The early SNA networks championed by IBM did this. At that time, computers were big, expensive, and scarce. IBM built a networking strategy that gave users on remote terminals access to a mainframe computer located at a different location. One of the first Internet applications was Telnet, which allows remote access from one computer to another. As technology and users evolved, the idea of remote commands execution (for example, rsh on Unix systems) emerged. rsh allows a command to be executed on a remote system; the results are then sent back to the local system the user is on. The idea of writing a program on one system and allowing it to call a procedure on a different machine also emerged: This become known as Remote Procedure Calls (RPCs) [1]. Data networks were invented to build distributed computing environments, and these environments have evolved along with the networks that provide the underlying services they need.

Soap services are hailed as the solution to the nagging problem of integrating disparate systems, but the idea is hardly new. The vision behind soap services is old, but agreement by all major vendors to the same standard for transparent distributed computing has never happened before now, with all vendors behind soap services. Following are some early examples of distributed computing ideas:

CORBA is one attempt at a set of standards for distributed computing started around 1989.

CORBA is the acronym for Common Object Request Broker Architecture, OMG's open, vendor-independent architecture and infrastructure that computer applications use to work together over networks. **Using the standard protocol Internet Inter-Orb Protocol (IIOP), a CORBA-based program from any vendor, on almost any computer, operating system, programming language, and network, can interoperate with a CORBA-based program from the same or another vendor, on almost any other computer, operating system, programming language, and network.** *[3]*

In the bold text, try replacing "CORBA" with "soap services" and "IIOP" with "HTTP, SOAP, and XML"; the sentence now reads as if one is describing a Web-based service. Similar to how soap services utilize a special language (WSDL) to define the interface for a remote piece of code logic, CORBA uses a language called Interface Definition Language (IDL) for the same function. The idea is the same: Define the interface unambiguously such that a computer can generate the correct request for remote services, and understand the result of invoking the distributed computation. But, because all vendors did not adopt CORBA, it did not live up to its hype.

Microsoft

DCOM [4] was Microsoft's answer to distributed computing. It is functionally similar to CORBA, and both are based on the RPC paradigm. At a technical level, DCOM is different from CORBA, but the differences are not important to the overall framework of DCOM because its high-level goal is similar to CORBA — that is, connecting completely different systems. As expected, DCOM favors a Windows environment, but it does work on other systems. DCOM, although popular with Microsoft users, did not catch on with all other vendors.

Microsoft is now supporting soap services, which it defines as follows [5]:

*A Web Service is a unit of application logic providing data and services to other applications. Applications access Web Services via ubiquitous Web protocols and data formats such as HTTP, XML, and SOAP, **with no need to worry about how each Web Service is implemented**. Web Services combine the best aspects of component-based development and the Web, and are a cornerstone of the Microsoft .NET programming model.*

Microsoft believes developers need a choice of languages, but not a choice of platforms. Microsoft insists on servers running its Windows software. Microsoft has built a snazzy development environment, complete with GUIs and drop-and-drag programming. It allows for building basic soap services without needing a sophisticated programmer. The architecture Microsoft has designed is language independent because it allows third parties to implement compilers for any language. This language-independent environment has value for several reasons: leverage of your current expertise, reuse of existing code, and using the best language for the task. Unfortunately, .NET insists on the Windows operating system. This is very constraining, as one can't leverage current expertise if it's not Windows, nor can one use Unix, Linux, or other operating systems. Finally, .NET limits the choice of hardware to smaller Intel-based systems.

Language independence is very important, but platform dependence is very constraining, which implies that Microsoft is half right.

Sun

Remote Method Invocation (RMI) [6] was Sun's entry into distributed computing. It is tightly coupled to the Java programming language because both client and server must be written in Java. It's a nice, general distributed environment allowing a Java program to invoke a method on another network-connected computer that is running a Java Virtual Machine (VM). Sun's entry is perhaps the weakest because of its tight coupling to the Java language; DCOM and CORBA, on the other hand, are language independent. RMI works well in the Java environment: It is very network friendly, and it handles the details of data and code serialization for transport across networks. If the only programming language were Java, this would be a very attractive solution. Java, while popular, will never be the language of choice for all applications. The biggest problem with RMI is its tight coupling with the Java language.

Sun now believes in soap services as the standard way to integrate IT systems, which it defines as follows [5]:

> . . . *an application that exists in a distributed environment, such as the Internet.* **A Web service accepts a request, performs its function based on the request, and returns a response**. *The request and the response can be part of the same operation, or they can occur separately, in which case the consumer does not need to wait for a response. Both the request and the response usually take the form of XML, a portable data-interchange format, and are delivered over a wire protocol, such as HTTP.*

Sun believes in platform independence for implementing soap services — as long as you use the Java language to implement this service. This is the write-once-run-anywhere philosophy of Java: As long as the machine has the Java Virtual Machine (VM), your Java application runs fine (well, maybe). The idea is to build a soap service on one system and then port it to another platform, with little or no change to the software of the soap service, which works sometimes, but not always. Sun recommends using a Sun server, but its standards don't insist on it (a good thing for users and vendors). Unfortunately, Sun is not open-minded about what language soap services should be implemented in — Java is the only choice Sun is willing to accept. Sun believes that Java is the best language for Web-based services — no matter what the soap service is — a claim that seems hard to believe and harder to prove. The reason for Sun's strategy is simple to understand because Sun owns the intellectual property rights to Java. Sun is half right; you should have a choice of platforms, but you should also have a choice of language.

As expected, Sun's system is typical of Unix development environments — it consists of many tools, with many options. This type of environment is flexible — you can do just about anything. Unless you are a guru, it is not easy to figure out how to do what you want. Unix experts will like Sun's approach; they can still use Emacs, and once they know soap service development tools, software is quickly implemented. Unfortunately, for nonexperts, the learning curve is long and hard. Unix and Java together are a powerful combination, but becoming an expert is difficult. Developers who like Unix and Java will find Sun's soap services development environment easy to work with as well as highly effective.

Table 12.1 highlights the similarities and differences between soap services and previous distributed computing environments.

Why Soap Services Are Better

CORBA, DCOM, and RMI are based on the RPC paradigm with tight coupling between what the client sends and what the server expects. The type and order of passed parameters are rigorously enforced because the parameters are marshaled and unmarshaled (the standard computer language for sending parameters across a network). This is a tighter coupling than required with soap services because soap services allow both an RPC and message paradigm (see the discussion about SOAP in this chapter). The RPC style of soap services is a mapping of the RPC paradigm onto the architecture of soap services: placing the XML encoding of the parameters into a SOAP envelope. The message-passing model is based on the exchange of messages and is far more flexible because of its looser coupling between client and server. Previous generations of distributed computation environments did not display the flexibility that soap services do.

Table 12.1 Distributed Computing Summary

METHOD	VENDOR	RPC	MESSAGING	ACCEPTANCES
CORBA	Consortium	X		Limited
DCOM	Microsoft	X		Limited
RMI	Sun	X		Limited
Soap services	Consortium	X	X	Everybody

Many big players in the pre-soap service days are leading the push to soap services. A vendor's soap services development environment is aligned with its history regarding other distributed environments it has supported. The vendors agree on the big picture of soap services; however, they have different ideas about how soap services should be implemented. The biggest players have business models that play to their strategic business advantages: Microsoft [9] believes in using the Windows platform, while Sun [10] is focused on the Java language. There are also other, less restrictive environments, such as Axis from the Apache Group [10]. Each system has advantages and disadvantages — the best choice depends on the particular attributes of your organization and what you want to do. The major vendors agree about what counts the most — it's not how you build soap services, but rather, it's what users can do with these services that creates value. Soap services have emerged as the best choice from the many different experiments in the realm of distributing computing environments. For the first time, industry agrees on a common method for access to remote resources across heterogeneous networks and systems. This is a good example of how learning from many generations eventually leads to a solution acceptable to most of the players — not an easy feat.

One powerful attribute of soap services that encourages innovation is the loose coupling between clients and servers. By agreeing on standards such as SOAP and XML, transferring data between all heterogeneous systems becomes easy. These standards explain how typed data is exchanged between systems that disagree about how data is represented. This loose coupling implies easy replacement of one soap service for another, if the defined interfaces of both soap services are identical. This loose coupling between the client and server with soap services gives consumers and developers tremendous flexibility in building and evolving these services because it promotes experimentation.

Why Implementation Platforms Don't Matter

Users and developers have very different views about how soap services are implemented: Users don't care, developers do. If you are developing soap services, choosing the development environment is a complex and difficult decision to make, but if you are the consumer of these services, you don't care at all. As discussed in this chapter, each vendor's solution has pluses and minuses; other open source solutions don't have as many features; there is no clear path to the future. Developers care about the total cost of development, which is dependent on the development environment. Different firms will find varying solutions that are the most cost-effective for them, depending on their in-house expertise. For example, financial organizations with strong Unix backgrounds will find Sun's J2EE and Unix/Linux the preferred platform on which to build soap services, while other firms without Unix skills will find Microsoft .NET the better solution. There is no best development environment because the choice is contingent on the organization's capacities and goals.

Most debate about soap services is about the best development and hosting platform, not the key points that make soap services valuable. Remember that the importance of soap services depends on who can use them and what the users do with them. Soap service clients are independent of the server platform and depend only on implementation of accepted protocols (XML, SOAP, . . .), thus the value to the client is independent of how the service is performed. The future looks bright for soap services because the vendors disagree only about implementation details, and users don't care about these details because clients can consume services regardless of the platform on which a service is hosted.

Users of soap services care only about adherence to the standard, performance of the service, and quality of the data from the service — not the server platform. The implementation details of soap services are unimportant to the consumer because the loose coupling between client and server hides how they work. The architecture of an organization's soap services is independent of the choice of development environment and implementation platform, which implies that as far as users go, the choice of development environments is not the critical factor — the design of the system is.

The architecture of an organization's soap services is related to the value these services have to the firm because this architecture is what determines the effectiveness of a firm's integration of their business logic. The degree of modularity these services exhibit (as explained later in this chapter) is one main factor in this value [11]. Fine-grained modularity provides more value because soap service consumers are able to pick and choose the services they need at a finer level. Firms sharing information most efficiently with those needing it have a strategic advantage over firms with less efficient information distribution. This ability to disseminate information quickly and efficiently is related to the architecture of the soap services.

Before going into a more detailed discussion of the value of soap service architecture, the reader needs a better understanding of the standards used to build soap services and how they fit together.

SOAP Service Tutorial

This section is a primer on the technology of soap services. Its goal is not to teach how to create soap services, but rather, to help the reader understand enough of the framework to apply soap service technology when solving business problems. This tutorial will help in understanding the next section describing the value of modularity and interchangeability of soap services. It introduces Extensible Markup Language (XML), which is fast becoming the language of the Internet, and Simple Object Access Protocol (SOAP), which holds the XML-encoded content. It explores how the interface of a soap service is defined with Web Service Description Language (WSDL), and how these interface definitions are registered, which allows users to discover them with Universal Description, Discovery, and Integration (UDDI). This combination of XML, SOAP, WSDL, and UDDI is a powerful set of tools that gives users tremendous flexibility in the services they can specify.

An analogy between a paper-based inventory check and the electronic soap service version is presented next. Suppose company A is interested in knowing if company B has 100 widgets. To accomplish this goal, someone at company A might print out a request (asking if company B has 100 widgets), place this request in an envelope, address the envelope, and then mail it. When this envelope arrives at company B, it is opened, and the content is removed and read. Next, a response is printed out and placed in an envelope, which is sent to company A. When the envelope arrives at company A, it is opened and the answer read. This simple process illustrates the intuitive idea behind how a soap service would perform this task. Table 12.2 maps these actions into their equivalent action as a soap service.

Table 12.2 Intuitive View of Soap Services

PAPER BASED	SOAP SERVICE BASED
Print out the request for checking inventory.	Create an XML document with the correct structure.
Place the request in an envelope.	Place the XML document in a SOAP envelope.
Send the envelope to company B.	Use HTTP to send company B the SOAP envelope.
A person opens and looks at the inventory request.	A computer opens the SOAP envelope and processes the inventory request.
A person creates a response to the request.	A computer generates an XML answer to the inventory request.
The answer is placed in an envelope and sent back to company A.	The XML answer is placed inside a SOAP envelope and sent to company A.
The envelope is received by company A and opened, and the answer is read.	The SOAP envelope is opened, and the XML document is removed and processed by the computer.

In the preceding example, this soap service is a paperless, electronic version of a business transaction reporting on the inventory of a particular item, as illustrated in Figure 12.2. The XML document is analogous to the paper request, as is the SOAP envelope to a paper envelope. The SOAP envelope contains meta information about this XML message.

XML — Extensible Markup Language

As it becomes the most popular way to encode data in a system-independent format, XML is becoming the lingo of the Internet. XML, because of its flexibility and extensibility, is able to encode just about any type of data imaginable. Interestingly, many soap service protocols such as SOAP and WSDL are defined using XML. Many vendors support XML, including most database vendors, and provide XML translation to and from their proprietary data format. It is this vendor acceptance that is propelling XML to its lofty role as the language of the Internet.

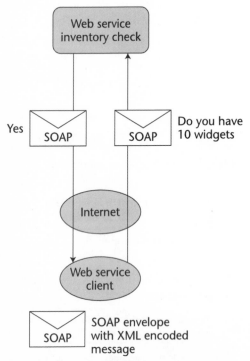

Figure 12.2 Inventory soap service.

A language such as XML is needed to define data and its associated structure because human languages are not specific enough to define data unambiguously. On a hot day, if I tell a student I think his shirt is "cool," what do I mean? Do I like the shirt, or do I think it will be cool on such a hot day? Even in face-to-face communications, the English language is not always precise enough to avoid misunderstanding, which is why a precise data description language such as XML is required.

Languages such as XML [12] are called markup languages because information is enclosed between tags that define the context of this data. The idea of adding meta information (meta information is data about the information) that explains, but is not part of, the data is not new. A complex specification called Standard Generalized Markup Language (SGML) existed in the mid-80s. It provided the framework to define HTML in the early 90s and, finally, XML by 1998 [10].

This meta information defines attributes about the data such as how to display it or what the data represents. Use of XML is split between two main applications: formatting data for consumption by people or formatting data for consumption by computers. Document-centric XML describes how data is structured for display. It includes tags such as <b> . . . </b> to bold a piece of text as it is displayed to a monitor or printed. The other main use of XML is data-centric XML, which describes structured information from databases, data structures used in computer programs, and other sources of complex data. Most data-centric XML documents are designed to be parsed by a computer. The data-centric use of XML, which places data in the context of what it means, is the base technology behind soap services.

In XML (and HTML), data is enclosed between tags that define the context of the data. Because tags can be nested, hierarchical complex structured data representation is promoted. What follows is a simple example of a data-centric XML document describing information about this book.

```
<book>
    <title>Network Services Investment Guide</title>
    <author>Mark Gaynor</author>
        <email>mgaynor@bu.edu</email>
    <publisher>Wiley</publisher>
    <ISBN>0471-21475-2</ISBN>
</book>
```

This example is one representation of this information. Note the hierarchical nature of this data: Each book has at least one author, and associated with each author is contact information. One nice (but inefficient) attribute of XML is the readability of its ASCII encoding. This simple example illustrates how information about this book might be encoded with XML.

There are two things that a program must know about the data it is processing: how this data is represented and what this data is. In the preceding XML example, the author is Mark Gaynor, but how is this information represented? In this case, it's not hard to deduce that this data is represented as a string; however, there needs to be a precise way to describe this to prevent misunderstanding about how data is typed. XML allows one XML document to be linked to another, called a schema, which defines precisely how the data in the associated data document is typed. To see how this works, consider an example of a soap service checking the inventory for some number of a particular item. The data this soap service needs is the code for the item and how many items to check for. One choice is to make the item code a string and the number of items to check for an integer (assuming you can check only for whole numbers of items). The code that follows is a simplified XML representation of this information for a soap service called check_item:

```
<xsd:schema xmlns="check_item"
            xmlns:xsd"http://www.w3.org/2001/XMLSchema"
            targetNamespace="check_item"

    <xsd:simpleType>
    <xsd:attribute name="item_code" use="required">
      <xsd:simpleType>
          <xsd:restriction base="xsd:string">
          <xsd:pattern value="[A-Z]-\d{5}"
        </xsd:restriction>
      </simpleType>
    <xsd:attribute name="item_num" use="required">
      <xsd:simpleType>
          <xsd:restriction base="xsd:integer">
            <xsd:minExclusive value="0"/>
            <xsd:maxExdlusive value="1000"/>
          </xsd:restriction>
        </simpleType>
```

This looks complex because it is; however, the important information for this chapter (and for your understanding) is in **bold**. This data the soap service receives is composed of two elements: item_code and item_num. This first element is a string representing the unique code of the item to check for inventory. This string consists of a single alpha character (between A and Z), followed by a hyphen, and then five numbers. For example, B-49899 is a valid item code, but B9-999 is not. The second param-eter is the number of items to check the inventory for. Its range is greater than zero and less than 1000. This example demonstrates the precise nature of how XML is used to describe data. In this case, the data is represented with no ambiguity; as long as you understand the rules of XML encoding, you can't interpret the data wrong.

The previous example of a schema is associated with an XML document that has data but no meta information about the type of data. Here is an example of how a piece of the XML request will look to the server when requesting inventory for 500 items of product D-23456:

```
<item_code="D-23456">

<Item_num="500">
```

This piece of code contains data that fits into the definition provided by the meta-data in the XML schema associated with this instance of its use. While the schema seems hard to read, its highly structured nature is easy for a computer to parse.

This very basic introduction to XML is enough to understand how soap services are put together, as well as why they have such potential to fun-damentally change business processes by putting information in the hands of those who need it. Don't be fooled into thinking XML is simple — it is not. Its flexibility to represent complex structured data creates complexity. Fortunately, this complexity is hidden from the users of soap services and not necessary to understand soap services at a business level.

SOAP – Simple Object Access Protocol

SOAP is the protocol used to exchange XML documents over the Internet. It provides the definition of XML-based encoded data used for exchanging structured and typed information between peers in a decentralized, distrib-uted environment [13]. Using XML, SOAP defines the messages passed back and forth between the soap service client and the server. SOAP is very flexible in how it defines the container for the XML data. This container is the SOAP envelope, which contains headers and a body. This SOAP mes-sage is not tied to any particular transport protocol and can be transported with HTTP, SMTP, or even FTP. SOAP is platform independent because any two operating systems with a SOAP stack can communicate via SOAP messages, allowing applications running on heterogeneous systems to speak the same language. SOAP is part of a distributed computing environ-ment that is mostly invisible to the users. SOAP is the first such distributed messaging protocol accepted by most major vendors including Microsoft, IBM, and Sun.

SOAP provides support for many aspects of distributed computing. Some functions it performs are listed here:

- Defines the unit of communication — the SOAP envelope that contains all message headers and body

- Allows handling of errors by the SOAP fault mechanism

- Provides a scheme allowing extensibility, which implies that SOAP's evolution is not constrained

- Allows several paradigms of communications: the RPC, direct document-centric, and indirect document-centric approaches to pass SOAP messages between users

- Determines the protocol used to transport SOAP messages (HTTP, SMTP, or FTP)

A simplified version of the SOAP message asking my wife Gretchen to bring home some pasta for dinner tonight is shown here:

```
<?xml version="1.0"?>
<soap:Envelopxmlns:soap="http://schemas.xmlsoap.org/soap/envelop/">
    <soap:Header>
        <To>Gretchen</To>
        <From>Mark</From>
    </soap:Header>
    <soap:Body>
        Please pick up some pasta for dinner tonight
    </soap:Body>
</soap:Envelopxmlns>
```

This simple example illustrates the SOAP envelope with a header and message body. The header in this case is who the message is to and from. The body is the message asking my wife to bring home some pasta for dinner tonight.

Two SOAP Models – RPC and Message

SOAP allows two styles of distributed computing architecture that use XML to exchange data between users: RPC and messaging architecture. RPC is most familiar to programmers because it fits within a model many have used. RPC uses SOAP and XML to marshal and unmarshal the parameters when calling a remote object or procedure. It is similar to DCOM and CORBA in its function. The messaging model is a far more flexible style of distributed computing because there are more options about how they exchange and process information. Both of these styles have their respective advantages and disadvantages; the particular application dictates which architecture makes the most sense.

The goal of the RPC style of SOAP messaging is to make calling a procedure on a remote computer similar to calling a procedure on the local system. The intention of RPC technology is to make calling this remote procedure (located someplace on the network) as transparent as possible to the programmer. One nice feature about the RPC style of communication is that programmers don't need to worry about encoding and decoding parameters because these parameters are marshaled and unmarshaled automatically. There are drawbacks, though, to this RPC architecture. First, both client and server must be simultaneously running in order for the RPC call to succeed. This makes sense in the context of executing procedures, but it is not flexible for more general distributed computing. Next, this RPC structure is not robust to changes in the API for the data being passed back and forth — that is, changes in the data passed are likely to break the interface. The RPC scheme of sending SOAP messages is familiar to most programmers, and it has some advantages but many limitations.

In contrast is the message-oriented style of passing SOAP messages. In this paradigm, messages are passed between processes. This is accomplished by APIs such as send_message() or get_message(). One advantage of this message-oriented scheme is that the client and server don't need to be running at the same time. The client can send the server a message that is queued until the server is up. When the server receives the queued message from the client, it processes it and sends the reply to the client. Again, this message can be queued until the client is available to receive the message. The message style of communication allows direct (such as in RPC) and queued structure, which gives this style of using SOAP more flexibility than the RPC structure. This message structure is a looser coupling between client and server as they exchange data between themselves because the program itself must encode and decode the data, which allows more flexibility in processing this data. Changes in the order or type of data passed are less likely to break a well-designed message-based soap service. This message-passing architecture is more flexible, but it is more complex for the programmer to use than the RPC paradigm.

The preceding discussion illustrates the flexibility of SOAP because it can be utilized in many different ways. SOAP supports both synchronous (RPC and direct messaging) and asynchronous (queued messages) communications between users. The application of the end-2-end argument illustrates why letting applications decide what paradigm of distributed processing (RPC or messages) is more appropriate than forcing the application into a predefined model. The ideas behind SOAP have existed in previous distributed computation environments, but for the first time, the major players have agreed to use SOAP to enable communication between distributed computer systems. SOAP has a good chance of success because it's flexible and extendable and, perhaps most importantly, because there is agreement among the biggest vendors that SOAP is the standard to define messaging between systems.

WSDL — Web Service Description Language

We know how soap services communicate with XML-encoded data within SOAP envelopes, but how do these clients know how to use these soap services? That is the job of WSDL. WSDL is just another Interface Definition Language (IDL), with one big difference from the IDLs of CORBA, DCOM, and others — everybody agrees that WSDL is the standard to describe soap services. As expected, WSDL is defined with XML. WSDL describes everything needed to use the soap service, as well as what to expect back from the service. This includes information such as what protocol is used to transport the SOAP message (HTTP, SMTP, or FTP), how the data passed between client and server is structured, what it means, and the URI by which the service is accessed. Soap services can exist without a WSDL description, but without this description, clients can't figure out in an unambiguous manner how to access the desired soap service.

Describing the interface to a particular soap service is important to its successful adoption. The details of how this is specified are not important, but understanding the type of information required to use the soap service successfully is. Following is a list of some important information that needs to be documented:

- For each message between the client and the server, the structure of the data must be described. This function associates a data type with each item of data passed in a message between client and server. The types element is used for this function.

- Each message passed between the client and the server must be specified. This function names each message that has a types definition. The message element is used for this function.

- For each soap service, the interaction between the client and the server needs defining. Does the service expect a request-response, or is the interaction different? The portType element explains details of this interaction between client and server.

- The client must know what protocol to use to connect to the server. This protocol might be HTTP, SMTP, or FTP. The binding element performs this function.

- The client must know the location of the service so that it can be called. This address of the service is specified with the service element.

The WSDL description of a soap service contains all of the preceding information. This data is enough for a potential user of the service to invoke it. The data defines how to access the service, the format of all data sent and received, and the interaction between client and server. These functions provide a formalized way that is unambiguous in how it defines this information. One nice attribute of this type of formalized description is that a computer can parse this definition and generate a code stub to invoke it. This means people don't need to dig into the WSDL description of the service because they can let automatic tools handle this task.

WSDL is a very important part of soap services because it allows potential users of a service to evaluate the service and learn how to use it. Without a WSDL description of a soap service, the user must depend on verbose descriptions of the soap service, which make precise understanding difficult. Without this detailed and precise definition of soap services, it's difficult for programmers to invoke them correctly because misinterpretation is so easy with informal descriptions. A wrong understanding of a soap service interface means the user will not interface correctly to the soap service. This means that a lot of experimentation is required to get it right — which is very resource-intensive. In essence, WSDL transforms a soap service into a commodity that any client can easily utilize.

UDDI — Universal Description, Discovery, and Integration

We know how to define what a soap service does, but how do users discover what soap services are available for consumption? That's the job of UDDI — it allows a soap service to be registered so that users can find the WSDL of the soap service. UDDI allows the provider of a soap service to advertise it by providing a central directory service for publishing technical information about soap services. This central database of soap services is called a registry. UDDI is implemented as a soap service that allows manipulation of this registry containing definitions of soap services. Some people believe this is a necessary element for the widespread adoption of soap services [10]. Similar to other soap service protocols, the agreement between all the vendors is its most valuable attribute. The idea behind UDDI is to provide a public place for soap services to be cataloged.

The details of how UDDI works are not critical to understanding the value of soap services. After all, you don't need to know how directory services at the telephone company work to understand why the service is useful and how to use it. The UDDI registry is essentially a yellow pages for soap services and allows for lookup of services. Users find great value when they have many choices for a soap service, as the ideas in this book illustrate — it is the value of the best of many. Service providers find value in this because it allows them to become one of the choices users have; when market uncertainty is high, this is potentially worth a lot of money to the service provider with the best service. UDDI lowers the entry barriers for new providers wishing to enter the market because its service enables users to know what their choices are. The important point about UDDI is that it defines the infrastructure needed for users to discover what services they have access to.

Putting the Soap Service Puzzle Together

So far, I have presented the framework defining what a soap service is and how to use one. I started at the bottom by defining XML, which is the language used to describe technical aspects of soap services in a precise way. Then this chapter discussed how this XML information is packaged into a SOAP message that is independent of any vendor's proprietary architecture. WSDL is discussed as the standard describing how to use a particular soap service. Finally, UDDI defines the infrastructure to build registries of technical information about soap services. Together these standards allow a commoditization of soap services. These are the main puzzle pieces that define how soap services work and why they have such value.

Figure 12.3 illustrates how these pieces fit together. First, the user discovers all the possible soap services from the registry and how to use them, then picks the best soap service for their particular needs from many choices. The user now has details of how to invoke the desired service and can do so directly.

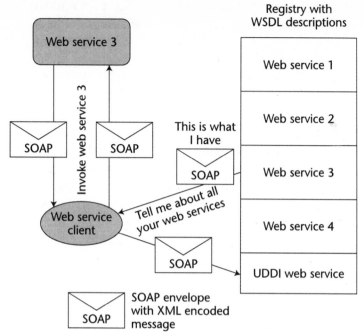

Figure 12.3   High-level Soap services architecture.

Value of Soap Services

Soap services fit well into this options-based framework because they allow users many choices. With all the big players supporting the standards behind soap services, users will have many choices because everybody is playing on a level field — it's possible to craft soap services that work on any client. Even more importantly, it's easy to copy the interface of any existing soap service and offer a competing version. This turns out to be great for users because many service providers can implement soap services that are interchangeable with each other, which gives users more choices. This is good for service providers because this standardization leads to lower barriers to entry. This plug-in ability of soap services creates value because the switching costs of changing services are small, which enables users to have more choices. My work illustrates the great value of soap services to both consumers and providers, as they promote users having many choices and allow successful service providers to make a lot of money.

Soap services force modularity because of their precisely defined interface, their distributed nature, and their loose coupling between client and server. You can't cheat with the modularity of a soap service because the loose coupling between client and server does not allow it. The only connection between a soap service and a client is an exchange of SOAP messages, making it impossible to break modularity rules. The implementations of soap services are hidden, which means clients can't take advantage of any implementation details. Because soap services are likely located on a remote computer, it is impossible to play tricks because the code is not located on the same machine. Modularity and soap services go hand in hand because the nature of soap services promotes and enforces modularity in design.

Value of Modularity

My approach to valuing modularity in soap services is similar to that used by Baldwin and Clark [11], which illustrates the value of modular design over its interconnected cousin in computer systems and is discussed in detail in Chapter 5. It allows keeping the best new module (perhaps picking the best outcome from many experiments) or keeping the old module, thus guaranteeing a higher expected value. To gain the most benefit from modularity, there should be many choices for modules that have the most potential to affect the total value of the system.

Figure 12.4(a) is an example that illustrates the value of modularity and choice as it relates to soap services. This is an example of a simple application: The online vendor is building a system allowing Web-based users to order products online. The vendor wants a system that is simple and inexpensive to design and implement but that will scale as his business grows. This is a common application: Sometimes a simple form is used, and other times a more complex shopping-cart approach is employed. For this example, we don't care about the front end, only about how the back end of the system works. This design based on soap services will be flexible now for easy experimentation, but scalable if the soap service becomes heavily used. This example will demonstrate the value of fine-grained modularity when initially designing the system.

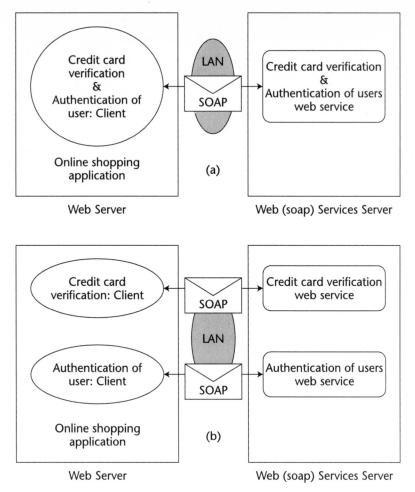

Figure 12.4   Web shopping application built with soap services.

The designer of a soap service decides on the degree of modularity — it can be coarse-grained, as in Figure 12.4 (a), or fine-grained, as illustrated in Figure 12.4(b). Two requirements of this Web-based shopping application are the authentication of users and verification of their credit cards. The top part of this figure has the authentication of the user and the verification of credit cards as a single soap service. This design works well and is very efficient, but it is not very flexible because of the link between authentication of the user and verification of the user's credit card for the given amount, which are unrelated functions. A valid user might be using an invalid credit card, and invalid users might have a valid credit card for the requested amount. If the designer changes one part of this system, the other part might be affected because of the interconnected architecture. This design makes it impossible to experiment independently with each function. In contrast to this efficient but inflexible interconnected design is the modular design in Figure 12.4(b). This design consumes two soap services: one to authenticate the user, and one to verify the credit card. This design is more flexible because these two soap services are completely independent of each other. Experimentation with one service does not affect the other service. Because of this modularity, it is expected that the design of the modular architecture will be roughly 1.4 (the square root of 2) times the value of the interconnected design, as Chapter 5 explains. Soap services give designers the flexibility to choose the degree of granularity of the design's modularity for the applications they build.

The idea of modularity in Figure 12.4(b) is very powerful because it allows this particular vendor to build these services in-house or outsource them. At first, it might make sense to implement these in-house with simple algorithms. Suppose, though, that there are problems with customer fraud and bad credit cards. It might pay to use a service provider that has a higher success rate of detecting problems because it specializes in this area. Soap services allow switching between in-house and outsourced solutions with almost no switching costs. This ability to swap one soap service for another is illustrated in Figure 12.5. For these tasks, the vendor calls one of several service providers that authenticate users and one of several services that check a credit card to verify its number and its current status. This figure illustrates several choices for each of these soap services. Assuming these soap services conform to the same published WSDL scheme, they can be swapped in and out at will — the switching cost is near zero because only the location of the service is changed. This ability creates an option value because the vendor can experiment with the different services and pick the best one (choosing between in-house or independently provided soap services) for his or her needs. This value is the best of many experiments discussed in Chapter 6.

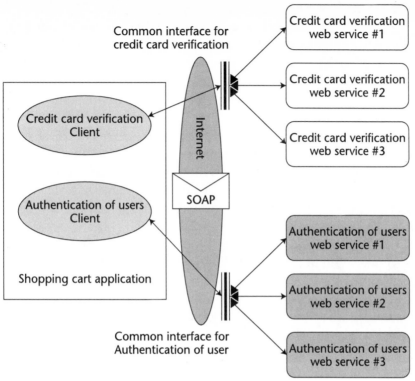

Figure 12.5 Plug-and-play attribute of soap services.

The degree of modularity should depend on the market uncertainty and the performance demands of the soap service. For soap services that are heavily used, it can be argued that breaking up applications into soap services with fine granularity makes meeting performance requirements for these soap services difficult, maybe even impossible. Highly successful soap services imply a more certain market. This means that the value of experimentation is low because the market is already well served. The advantage of efficiency by combining soap services into coarse granularity modules (as Figure 12.4(a) illustrates) is more important than the ability to experiment with soap services from many different vendors. A likely successful strategy predicted by the theories from Part One is starting off with a high degree of modularity. Then, as the market evolves and your understanding of this market grows, combine modules as performance demands. This strategy creates value because it captures the value of innovation when market uncertainty is high and the value of efficiency when market uncertainty is low.

There is great value to organizations whose IT infrastructure is built with fine-grained (basic) soap services. Architecture based on the building blocks of basic soap services creates value in terms of the internal integration of different IT systems and the benefits of linking with external partners. Soap services are the best of all worlds because they allow both internal and external integration of heterogeneous IT systems. Soap services can link IT systems within an organization, and the same technology extends this integration to external partners. This architecture encourages in-house development of services and, when desired, extends this integration past the organizational boundaries to include business partners. For now, soap services are the design methodology offering the most options down the road because of this ability to integrate both a company's internal IT systems and its external business partners — in short, they create value by increasing the number of choices users have.

Conclusion

In this chapter I discussed the classification of network-based applications and services in the context of what a Web-based application and service are. I explained how Web-based applications/services have attributes of both the distributed and the centralized management structure. They offer a distributed management structure that allows users to have choices or a centralized management architecture that has many business and technical advantages. This flexibility has value to both service providers and users, as the success of Web applications has demonstrated.

This chapter discussed the newest, hottest technology: soap services. At the risk of sounding overly optimistic, I believe that soap services might create tremendous value within organizations that use them to build IT systems. Soap services allow business logic to be exposed to anybody you want, in a format that is easy to understand. The technology includes XML, SOAP, WSDL, and UDDI. These ideas are not new, but agreement about how to offer distributed computer services over networks has not converged before the technology of soap services emerged. Vendors and service providers finally realized the value of what users have always desired — the ability to access services over a network with whatever device they happen to have. The jury is still out, but a lot of big companies, such as Microsoft, Sun, IBM, and others, have bet most of or the entire farm on this new technology. The theories of this book imply that soap services are likely to be successful because they are flexible in the management style they allow, and promote modularity, thereby giving users many choices.

The theories here fit well with how these new technologies are evolving. In building VoIP applications, SIP is becoming successful because of the flexibility it allows in the choice of management structure. 802.11 is stunningly successful, and it also allows a continuum from distributed to centralized management. Finally, soap services and Web applications are becoming successful and exhibit the important trait of flexibility in the choice of management structure. As my ideas suggest, when market uncertainty is high and dynamic, flexibility becomes more valuable. These examples illustrate the breadth of application to which the ideas from Part One apply.

The last chapter in this book discusses the generality of theories from Part One, examines a common thread found in all the cases, and gives advice to different audiences about how to apply the theories in this book to different environments. The chapter starts by discussing the vast differences between the Internet and PSTN, yet noting their similarities in the context of how they evolved to give users choices in management architecture. The evidence for this argument comes from Chapter 3 (end-2-end), Chapter 8 (email case study), and Chapter 9 (voice case study). Next, it discusses the common thread found with all the cases. The common attribute of most successful services is allowing flexibility in the choice of management structure. The final chapter ends with a section about how different readers should apply the ideas of this book. It discusses how investors should think about ventures related to network-based services. It illustrates that uncertainty is convertible into value when the network infrastructure allows experimentation. It will help investors understand what management structure works within the context of their investments and how they should help guide the direction of these ventures. It discusses how managers can take advantage of uncertainty by allowing a degree of flexibility tuned to the uncertainty of the market. This last chapter aims to focus on the particular decisions faced by professionals in the financial and Information Technology fields.

CHAPTER 13

Conclusion and Advice to Readers

This final chapter ties everything together by first discussing the generality of the arguments presented in this book. Next, I review the common thread found in the case studies from Part Two and the predictions in Part Three. The chapter and book ends with a short section discussing how different types of readers (for example, investors, service providers, IT managers, consultants, and individuals) should think about the theories presented in this book. Each of these readers will apply the theories in their own ways, according to their job responsibilities and interests. This final chapter explains the power and flexibility of the theories and how different professionals will apply the ideas presented in previous chapters.

Generality of Argument

The Internet and the PSTN, which had different starting points, are evolving into similar structures. The Internet started out with very little in the way of internal network-based services, but over time, it has evolved into a network with more core-based services (such as Hotmail, Yahoo!, and CNN) and middleware (such as Public Key Infrastructure and QoS). The phone network did the opposite; it started out with no intelligence at the edges, but evolved into a network with more sophisticated Customer

Premise Equipment CPE (PBXs and smart phones). ISDN even permits customer configuration of internal network-based services. These two examples demonstrate that two very different networks have evolved into a similar two-tiered structure, allowing innovation at the edges and migration of successful innovations inside the network. This gives users and service providers the benefits of a centralized management structure. Two different networks evolved into a similar two-tiered structure, which illustrates the power of this argument.

This structure allows features of services to incubate in the outer region (see Figure 7.9), after which the successful features are offered by the network. Both email and voice services have this organization allowing feature migration. Email has seen features such as address format, message transport, and message encoding (MIME) migrate from mere experiments to highly centralized mail servers. PBX features, such as caller ID, generic PBX features, voice mail, and ACDs, have migrated from the edges of the network to the core. The fact that two fundamentally different networks have evolved a simple mechanism (which allows services and service features to migrate into more centralized managed structures) shows the generality of this idea.

This argument is general and applicable to other situations where a choice exists between environments that allow experimentation and environments that are not flexible, yet offer other advantages. One is in the upcoming business-to-business Internet companies. The idea is to create services that allow efficient business transactions among one another. One important question to consider is whether a b-2-b service should be centralized or distributed. Centralized b-2-b models offer a consistent and effective way for companies to cut transaction costs, but they are inflexible to individual business needs. Distributed b-2-b services are more flexible and can meet particular needs better, but they don't offer the efficiency or consistency of centralized systems. As my thesis suggests, the uncertainty of what these businesses want and the ability of providers to predict what will work are critical factors in determining what structure to build.

One example of this occurs in the automobile industry[1] and how b-2-b services have evolved within it. In this industry, there are several levels, each with distinct uncertainty. There are parts for automobiles — these have low uncertainty. They are a commodity, and users don't need choice. For transactions involving parts, centralized systems have evolved. In other areas, such as service agreements, where market uncertainty is high, centralized services have not evolved because distributed systems are more successful because of their flexibility. This example shows how my

[1] In a conversation in the defense of this thesis with Marco Iansiti.

argument of the value of experimentation, which provides choices for users, depends on the uncertainty.

While useful and general, this argument in its current stage has limitations. Foremost is the difficulty of input data, such as market uncertainty, or the difficulty of arriving at a realistic value for the business and technical advantages of centralized management for a particular situation. The usefulness of this book is a high-level view about the choice of management structure in light of the ability of service providers to know what users want. It cannot make precise forecasts; rather, it predicts trends.

Another weakness of this work is the complexity of each case. While applicable to a wide range of situations, each case must be carefully analyzed. There are many subtleties when management structure shifts, and each case is different. This was true in the two case studies. In email, technology was potentially important in the shift to Web-based email, whereas regulation was unimportant. In the case of voice services, regulation needed careful treatment, while technology was not a factor requiring thoughtful analysis.

Common Thread — Flexibility

The common thread in all the previous examples is the success of protocols and infrastructure that allow flexibility in management style. At all levels of the stack, many protocols and applications being adopted are flexible in this regard. At the bottom layer, the wireless link market seems to be tipping toward the 802.11 set of standards for LAN connectivity. As Chapter 11 illustrated, the wireless Wi-Fi link market is very flexible in the management style it allows — from a very distributed, do-it-yourself infrastructure to a more centralized structure, promoting more efficient resource usage, technical advantages, and easy billing. At the application protocol layer, SIP and Internet email protocols demonstrate the same type of management flexibility that the 802.11 infrastructure allows. As Chapter 10 demonstrates, SIP allows true end-2-end services or more centralized architecture, when services are designed around an SIP proxy. Internet email allows both a distributed and centralized management structure, as discussed in Chapter 8. Finally, at the highest level, are Web-based user applications, which have this property of flexible management structure. Web-based applications such as email can have a distributed structure, with ISPs running POP-based Web email, or a more centralized structure (Hotmail). Protocols that are flexible in regard to management structure and successful at many layers illustrate the value of flexibility in management architecture.

Link layer protocols send data directly from host to host on the physical media, which may be twisted pair, fiber, or wireless. My focus is on wireless — 802.11 and 3G. The current evidence indicates that 802.11 is growing fast in many different markets. The commodity nature of the equipment, easy setup, enthusiastic response from users, and good performance enable this growth. Fewer distributed models of 802.11 are becoming popular, as organizations such as Joltage and Boingo start to impose a more centralized management structure with wireless service in airports, hotels, and other hot spots. The future of 802.11 seems solid as a wireless LAN technology, and it might prove useful for other applications. The flexibility of management choice with 802.11 has stimulated the growth of this new technology because it meets the needs of so many users.

Higher up the protocol stack are application layer protocols for building services such as SIP and megaco/H.248, or the Internet set of email standards, along with the applications that use these protocols. SIP and megaco/H.248 are two protocols used for building Internet services with voice as an integral component. As discussed in Chapter 10, SIP is very flexible in the range of management structure it allows, but megaco/H.248 is not. This is similar to the flexibility with 802.11 because SIP and 802.11 both allow a continuum of management structure, meeting the needs of different user groups. SIP allows end users to experiment without anything within the network knowing what these end users are doing, and SIP allows centralized service providers to manage SIP proxies that impose a centralized structure. SIP is catching on because of its flexibility — it allows users to do what they please. Distributed management gives users freedom to experiment, while centralized management gives users efficient use of resources, technical advantages, and easy management. The email set of standards also exhibits this trait of flexibility in management structure. The protocols allow building email systems with a centralized or distributed management structure. End users can run Sendmail on their local machine, or they can use a more centralized server along with the POP protocol or a Web-based centrally managed email service such as Hotmail. Internet email users have choices in how their messages are managed: With POP, the user must manage messages on his or her computer; however, with IMAP and some Web-based email services, users have the choice of leaving email messages on the email server, thus delegating the responsibility of message management to this server — a more centralized management structure. SIP and the Internet mail set of standards are two examples demonstrating how the same protocols are able to build applications with both centralized and distributed management structures.

This common thread is seen with voice services because users have a choice between a distributed managed PBX and a centrally managed Centrex. The PSTN has evolved into a structure that gives users choices on how they want to structure their network management architecture. Even prior to deregulation and user choice in equipment and service provider for voice services, users could choose between Centrex and PBXs. User choice was limited, though, because AT&T was the vendor for both the PBX and the Centrex service. After deregulation, there was true competition between PBXs and Centrex, as well as competition between different PBX vendors, which gave users many choices. Similar to successful Internet infrastructure and applications, voice services in the PSTN give users a choice between distributed and centralized management — the choice of management structure that is best depending on the market uncertainty. Centrex service provided by the large centralized carriers does well with basic services where market uncertainty is low, and the distributed management structure of PBXs works best when market uncertainty is high, such as with cutting-edge features like computer voice integration. Basic and advanced voice applications in the PSTN provide flexibility in the management structure they allow.

The common thread in both the PSTN and the Internet to creating the most value is flexibility in management structure. By allowing distributed management architecture when market uncertainty is high, the value of experimentation is captured. Allowing centralized management under conditions of lower market uncertainty enables efficient use of resources, which gives users the lowest price. Protocols, services, and applications that allow both centralized and distributed management structures meet users' needs independent of the market uncertainty, which creates the most value for users. It does not matter what the underlying network infrastructure is — users like flexibility in choice between distributed and centralized management architectures.

Advice to Readers

Investors

As I write this, the telecom industry is in a slump, mostly because it failed to meet expectations that made no sense in any world and massively overbuilt capacity. This industry is clearly strong in many ways: Consider its vast revenue. I consider myself a basic user of telecom services: I pay for a home phone (most people do), my family has two cell phones with way more minutes than any of us use (not uncommon), and from the cable company we have advanced basic (no premium channels such as HBO or Showtime) and broadband service (not so common yet). I am paying more than $200.00 per month for these services in a package that many middle class users now have — this is a lot of money.

VCs

The venture capital community is vital to finding the best services because they finance much of the experimentation required to find good services in uncertain markets. The theory suggests that there are many winners and losers with this experimentation, which implies that VCs must have intelligent strategies that consider the market uncertainty in the context of deciding how to structure their investment portfolio. VCs take tremendous risks and need to ensure that they manage the risk by applying an options framework that maximizes their value because VCs need to generate the large capital to fund the services of the future. Understanding these theories and how they apply to funding companies in the network service industry is a strategic advantage in today's uncertain world.

Applying the framework from this book to funding startup companies in the networking industry should affect the number of ventures versus the size of the investment in each deal. When investing in highly uncertain markets, VCs should consider as many small ventures as possible. Each deal is an experiment with a high degree of uncertainty. One of these might be that great service or product everybody is looking for and will make lots of money. If the potential market size is large enough, betting on many services or products, each with a relatively small investment, is likely to maximize value because it takes only one big success to offset the many small investments that did not pan out. High market uncertainty means VCs will lose many of their bets, but with research and luck, the likelihood of tremendous success is reasonable and worth risking. When market uncertainty is low, it makes more sense to invest in fewer, but larger, deals. Low market uncertainty implies that the strategy of spreading out your investments in many small pieces is not optimal because it is unlikely that the companies you are investing in will hit on a new idea, and because it is more likely that everybody playing in the certain market will be able to deliver services or products that meet users demands. This means the payback on investment won't be stellar, but it will be a fair return for a commodity market. In this case innovation takes a back seat to efficiency, which implies that investing in larger companies that have the advantages of a centralized structure is the best strategy. By considering market uncertainty as a factor, the mix of a VC's investment portfolio will increase the expected gain.

The way VCs fund ventures in stages fits my theory well. When market uncertainty is high (the first round of funding), there are many small deals. Then, as some ventures don't pan out, a few successful companies get more funding. At first, when market uncertainty is high, VCs make many smaller bets. When market uncertainty is low, VCs make fewer, but larger investments. When market uncertainty is low, VCs don't need many deals to capture value. This is the time to take advantage of centralized management and efficiency and not worry about meeting the market because it is so easy to predict users' preferences.

VCs should take advantage of experimenting in the most lucrative region (shown in Figure 11.7) by investing in small startups that have expertise in their particular area. VCs are putting money into firms that are experts in their particular area, which means that a VC can maximize its gain by investing in several firms that are expert in their particular area and the most productive with experimentation. Good ideas are expected to come from the innovative startups, not the big established vendors and service providers. It makes sense to provide funding to those most likely to have the best ideas. This strategy of experimentation in the critical area where value is maximized is what VCs need to maximize their value.

Service Providers

Service providers can be huge with vast centralized infrastructures (RBOCs, long distance providers, and wireless carriers), or they can be niche companies serving a small group with a more decentralized management structure. Each type of service provider needs different strategies to maximize the value of providing services to users. Large centralized organizations are expected to be the most efficient, but also the least innovative. Smaller and more nimble service providers are more likely to have better ideas, but they have a hard time efficiently providing these cutting-edge services. With the correct strategy, either can win because there is room for small and large service providers because different users have different needs, as seen throughout this book. Following the arguments in Chapter 11 about the coexistence of distributed Wi-Fi and centralized cellular wireless infrastructure, small and large centralized services can add value to each.

Large Centralized Providers

Companies such as Sprint and the RBOCs are experienced in providing network services within a centralized management structure. As Chapters 2 and 4 explained, this type of centralized management structure has many advantages. The telephone network illustrates the success of this type of centralized architecture. Unfortunately, these advantages of centralized structure make it less likely to think "out of the box." It is "out of the box" thinking, though, that is needed to find great services in uncertain markets. These big service providers have a dilemma: They are in the best position to provide many types of services as long as the number of users is large, but they are the least likely to figure out the services that will be successful enough to merit the efficiency of a centralized management structure.

The strategy for these large carrier-type service providers is to do what they do best — provide services that are known to be successful. In most areas (see Figures 11.6 and 11.7) where these large organizations don't have cutting-edge dexterity, their experimentation is unlikely to be successful. By funding or buying ideas from smaller niche service providers that have specialized expertise, these large organizations can discover a stream of services that migrate into their centralized structure, providing a continued flow of revenue. This is an options value — the option of capitalizing on the innovation of others. Big service providers don't need to have the best ideas; rather, they just need to watch and see what the ideas are, letting the market select good candidates for migration into their centralized umbrella. By doing what they do best (efficiently providing services to many), these large centralized service providers can be successful.

Centralized service providers supplying a basic infrastructure such as basic IP (including wireless), second-generation cellular, and traditional telephone companies will also gain value if they open up their network to capture value from the innovation of others (see the open garden discussion in Chapter 11). Billing is one area in which these infrastructure providers can add value to the services offered by others, as DoCoMo's I-mode has demonstrated. Depending on the business model, centralized providers might gain revenue from additional data from services with which they have no affiliation. Encouraging other small service providers that are not affiliated with the infrastructure provider helps build a fertile environment that fosters innovation. These successful unaffiliated service providers are future partners to the infrastructure service providers. I believe that the options framework presented in this book implies that the open garden architecture described in Chapter 11 will maximize value.

Smaller Niche Providers

Smaller service providers have the opposite quandary of large centralized carriers: They are the most likely to find great services, but they don't have the centralized structure required to scale up these services efficiently to many users, should they become successful. These smaller niche service providers have a better chance of finding great services because their experimentation will be more focused (see Figure 11.6). These smaller niche service providers are more numerous, which implies an additional benefit from experimentation (as Chapter 6 shows) because of the greater number of experiments. The problem with these niche providers is that if they become too successful, the success can kill them. There are many examples of companies whose very success was fatal because their business model did not scale up. Having good ideas is only half the battle, as you will then need to provide a successful service efficiently to continue winning market share. Smaller niche providers do a great job of figuring out what works best, but they have difficulty scaling up when their services become popular.

These smaller niche service providers have two strategies for success: affiliation with large centralized providers or scaling up the organization as it becomes successful. Affiliation with large centralized providers has value to niche players that have rapid success. These bigger carrier-type providers have infrastructure to bill and manage customers, which is hard for the smaller service providers to match. The two most successful Web-based email systems (Hotmail and Yahoo!) both acquired this technology from smaller, more nimble innovative companies, as Chapter 8 explores. Efficient management is not the strong point of most startups: Being fast, creative, and nimble is. This style is the opposite of the one used by the bigger providers, which are efficient but not fast or creative. This dichotomy creates a synergy between smaller niche providers and larger centralized ones: They create value for each other. The other strategy is to go it alone, scaling up as needed. Companies such as MCI and Yahoo! have been successful with this strategy, but it's hard to do. In new markets where market uncertainty is high, small niche providers have tremendous business opportunities with either strategy they choose: affiliation or scaling up.

IT Managers

IT managers are responsible for providing services to others within their organization. Their success is linked to the happiness of their users. Sometimes when you can't figure out what users want it is best to let users have lots of freedom because they may be able to solve their own problems. Research universities are good examples of environments where user freedom is valuable. The most important faculty members are doing cutting-edge research — by definition, their needs are open-ended, with lots of uncertainty. Sometimes users' needs are so well understood that it's easy to satisfy them, which implies that centralized control can be imposed without losing much value. Large law firms where office applications are well defined are a good example of this type of organization because most users have similar needs. In this case, centralized management is likely to save resources and not constrain users because their needs fit into well-defined categories. The choices are complex, but understanding how market uncertainty affects the value of such choices will help maximize the return on investment.

CTO/CIO-Level Decisions

The main focus of this book is geared toward senior IT management. Chapter 4, on management structure, provided a framework senior managers can apply when deciding on high-level IT policy. Senior-level IT executives make the high-level strategy choices that guide the organization's IT plan. Questions including what services and applications are needed, as well as how to best provide them, need to be answered. Fundamental directions need to be established. Deciding how much expertise to build in-house versus what to outsource is critical to a company's success. The architecture of IT infrastructure and the policies governing its use can become a strategic advantage or, if bad decisions are made, a strategic liability. Managers that understand the link between market uncertainty and choice of management structure will craft an IT organization structure that is flexible to adapt to current conditions, which might mean centralized architecture for well-understood services or a more distributed structure when exploring new services with great uncertainty.

Individuals

The growth in network services for individuals is steady: Between basic phone services, advanced phone services (such as voice mail), wireless telephone, basic cable, premium cable (such as HBO), and Internet service providers, many individuals are paying a lot of money. We are faced with many choices; how should you choose among the options? Should you buy or rent equipment? Should you consolidate your services to a few large service providers? These answers depend on your personality and the market uncertainty. Do you need to have the latest, best, fastest? Or do you want stability and like to change technology only when you really need to?

Looking at how different users upgrade software on their personal systems illustrates the different personalities users have. Some users (like myself) need a reason to upgrade. I don't upgrade my software unless I really like a new feature or am forced to because of downward compatibility. I have upgraded Microsoft Word three times in about seven years. Once I needed to because of compatibility issues; I needed to exchange documents with others whose versions had incompatible features. The other two times I upgraded because Microsoft created features that I really liked: the real-time spelling and grammar checker and conversion of a Word file to HTML. My behavior is very different from that of one of my old office mates at Harvard. One day he upgraded all the office systems (including my system). He had no reason, except that it was the new version. For me, this was a disaster: I could no longer print my thesis out, and it was due soon. I was linking files in a complex way that was not consistent across these versions. To my office mate's credit, he did quickly undo his upgrade. This example shows how different users will adopt different strategies for managing their personal choices.

There are many different criteria for these personal decisions. When market uncertainty is low, the criteria might be secondary attributes, such as do you want that ugly answering machine around and do you have a good place for it. With low market uncertainty, decisions like this don't have much impact on the service because many solutions work for most users. When market uncertainty is high, the demanding user wants control and will pay extra for it. Different users will make different decisions based on the market uncertainty and their individual preferences.

A Formal Theory and Real Options-Based Model

Being able to visualize the trade-off between market uncertainty, the number of experiments, and the benefits of centralized management is a powerful advantage for today's investors and managers. The complex interrelationships among the factors influencing the choice of management structure are clearer when the components affecting the choice have a visual representation. The formal theory presented in this appendix is the framework for a mathematical model that illustrates these trade-offs. This model, based on real options, is a visualization tool to help illustrate how these different factors interrelate with each other.

This appendix focuses on the intuitive ideas presented in Chapter 7 and develops a formal theory about the value of network-based services based on how users make choices in uncertain markets and how much it is worth to give users these choices. Some of the text and figures from Chapter 7 are repeated here for continuity in reading. This theory is based on a set of assumptions about network-based services and how users adopt such services in uncertain markets. The theory starts out by examining the value of a network-based service that has only a single generation. The theory is expanded to account for how network-based services evolve over time and how both service providers and users learn from past generations of the service. This theory explains why market uncertainty increases the value of experimentation and choice and the importance of market uncertainty to the evolution of network-based services. It illustrates how services evolve from generation to generation as market uncertainty changes, based on the current conditions.

This appendix develops a mathematical model based on the theory presented in this appendix, linking the business concept of market uncertainty to the technical aspects of designing and implementing network-based services. The model illustrates the value of a network-based service to its provider as the service evolves in uncertain markets. Based on the real options theory and work by Baldwin and Clark [1], this model builds on previous papers by Bradner and myself that demonstrate how modularity in protocols increases value [2][3][4]. This model is a framework to aid in understanding the trade-offs among market uncertainty, the ability of the management structure to foster innovation, and the business and/or technical advantages of building services with a centralized management structure.

The model shows that when market uncertainty is high, a highly centralized management structure results in services that poorly match the market and often fail. Finding features wanted by users may be difficult with less flexible, centrally managed architectures. This model helps understand how and why the Internet has been so successful in creating services that meet market need, and it will enable current architects of the Internet to continue its successful evolution. In addition, the model sheds light on the implication of current important architectural fog, such as Network Address Translation (NATs) and firewalls, with regard to the cost of such devices in terms of lost innovation in network-based services.

The theory in this appendix agrees with the results of the two case studies in Chapters 8 and 9. Both voice and email services have similarities in the context of what management structures users favored and when. In both cases, users migrated to a more centralized architecture when market uncertainty was low. Users favored a distributed structure when market uncertainty was high because of the need for experimentation. Finding agreement with services from the stupid Internet and the smart PSTN is strong evidence in support of this theory and model.

The first section of this appendix promulgates this book's theory of network-based service architecture. First, the theory defines a set of assumptions to classify types of services and the conditions in which the services evolve. Then, it helps determine the value of a group of services, given that the services and the market for these services follow the assumptions. An explanation of one fundamental concept in real options, the value of the best of many experiments, follows. This theory forms a baseline of a mathematical model validating these ideas.

Theory

This theory provides a framework useful for analyzing what management structure works best for a network-based service. It does not provide absolute numbers or rules. It illustrates general relationships between market uncertainty and the power of parallel experimentation with market selection compared to the benefit of central management. This framework is useful for developing a strategy to maximize the expected return from investments when building new network-based services, by structuring the service management architecture to allow a proper amount of innovation in feature development, given the market uncertainty.

There are three stages in this theory, beginning with a few simple assumptions and an explanation of the value of providing a choice of services to users, then progressing to a more complete accounting of how services evolve over time in response to market pressures. The first stage shows that the value a provider of a service receives can be random due to the effects of market uncertainty. Next, I show that giving users many options to pick from provides the best chance of achieving a service with a superior market match. Essentially, if users' needs are unpredictable, then by giving them many choices, a really good fit with the market is likely. In the worst case, users will always be just as happy as if they had only one choice. In the second stage, the theory expands to account for management advantages gained from a centralized management structure for network-based services. I hypothesize that when the advantage of a more centralized management structure outweighs the benefit of many experiments, a centralized management structure may be justified. Both the first and second stages look at a single generation of a service; in stage three, the theory accounts for how services evolve from generation to generation. My hypothesis is that at each generation of a service, service providers learn from the current generation about what will work better for the next generation.

Stage 1 of my theory starts with several assumptions, definitions, and a basic rule about the value of users having many choices.

ASSUMPTION 1   The market demand for network-based services has a degree of uncertainty. This means that service providers may not accurately predict the value they will receive for providing a service because the value of the service to its provider contains an element of randomness. This market uncertainty is denoted as MU.

MU, as discussed in Chapter 6, is complex in that users' expectations evolve along with the technology [5]. Users don't know what they want or how their wants will change over time. The metric for MU should be independent of the market size because a successful service can alter current markets and create new ones.

There are many examples illustrating market uncertainty. Just consider that the Web was unpredicted or that PBX vendors believed they would capture the LAN data market. For years pundits believed the X.400 suite of email protocols would be adopted by most users, but instead, the Internet architecture became the industry standard. These examples show how wrong vendors and industry experts can be.

ASSUMPTION 2 Experimentation with services is possible. The artifact produced by a service experiment performed by a service provider[1] is a service instance. I define a service experiment as the development and deployment of a service instance. There exists a way to gauge the market success of each service instance.

DEFINITION 1 *X* is a random variable denoting the value to the service provider of a single instance of a service.

DEFINITION 2 A service group is a set of service instances, with each instance available to the user as a service. Users have the option of picking the service instance within a service group that best meets their needs.

DEFINITION 3 $X(i)$, $i = 1 \ldots n$, are random variables denoting the value to the service provider of providing the ith particular instance of a service within a service group of size *n*. With this simultaneous experimentation, each service instance does not benefit from the other instances of services within its service group because they occur at the same time. For the effects of sequential experimentation with services, see Assumption 8.

RULE 1 $E[Max(X(1), \ldots, X(n))] >= E(X)$, that is, the expected value of the maximum of *n* simultaneous attempts at providing service instances by some service provider may be far above the expected value. As *n* or *the market uncertainty* increases, the possibility of a truly outstanding market match grows.

The left side of this equation is the value obtained by the service provider's best matching service within the market. This particular service instance is the "winning" service. As the number of experiments increases, the expected value of the service with the best market match will grow at a decreasing rate.

[1]The user may be the service provider for services with architectures such as the end-2-end principle.

As MU increases, the value of the best of many services increases at a linear rate. This rule is a formalization of Figure 6.1 in Chapter 6.

There are many examples of how good the best of many experiments can be. Consider the Web itself; it was the best of many attempts to unlock the potential of the Internet, and it was an astonishing success. Alternatively, consider Web-based applications; eBay is stunning in its success, but other ventures, such as selling furniture online, failed. The ability to pick the best of many experiments can have tremendous value.

One way to view Rule 1 is in the context of the options theory; having a choice is analogous to having an option. This theory follows a fundamental idea in the options theory — choice and uncertainty increase value. To see this intuitive logic consider the following choice: Would you rather own an option for a single stock or own the option on many different securities (including the single stock) with the understanding that you can exercise any one option (but only one) of many in the option portfolio? Being able to pick the stock in the group that has gained the most value is clearly the more valuable choice. As the uncertainty about the stock's prices grows, so does the value of users having choices.

Giving users too many choices may have the undesirable effect of fragmenting the market. The more services users have to pick from, the smaller the value of a poorly matching service because of the number of better options the user has. More services within the market imply a greater range of outcomes as to how well any particular service instance will match the market. It is possible that many service providers will lose everything if their service is a poor market match and many better matches are available for users to choose. Another concern is that many parallel experiments are not an optimal solution in regards to society as a whole. The increased total cost of providing n services, while knowing that many of those services will not succeed (fortunately the world is not rational[2]), is not the lowest-cost solution. This view is countered by the greater value of the best service; the winner does have the optimal solution and will profit handsomely. In general, the expectation is that the market percentage captured by a service instance is proportional to how well the instance of this particular service matches the market.

Next I present a set of stronger assumptions, leading the way to a deeper theory about the type of service management structure that works best for a given degree of market uncertainty. The following helps to clarify what it means to allow easy experimentation in a network:

[2] Just look at the changing market value of beanie babies™, Pokemon cards™, and dot-com companies.

ASSUMPTION 3   The function representing the value to the service provider of providing a particular instance of a service that best matches a particular market is nonlinear. More experimentation and greater uncertainty increase the expected value.

That is, by undertaking more experiments when MU is high, the expected value of the best of these experiments might far exceed the value of a single attempt at providing the service. The variability of the service's value determines the value of the option [6].

ASSUMPTION 4   Experimentation with providing service not requiring changes within the network infrastructure or permission from the network manager (for example, true end-2-end services) is, in general, less disruptive to other users and less expensive than experimenting with services that require changes within the network or permission from a central authority.

Assumption 4 is very important to innovation. Changes within the network infrastructure require permission from those controlling the network. New end-2-end services do not require permission. For example, one person can implement a new HTTP header without asking. Then, by proposing it to the IETF, the market has the chance to accept or reject the change.[3] If Tim Berners-Lee, the Web's creator, required permission from a network authority to experiment with the Web, it is less likely that he would have been successful.

ASSUMPTION 5   If a real or potential market exists that is not being met, then the less disruptive and less expensive it is to experiment, the more experiments there will be.

One good example of this is the large amount of experimentation with Web-based applications. Clearly, there is strong market demand for some Web applications. It is also easy to experiment with different Web services because of the open nature of Internet and Web standards and the distributed management architecture of the Internet. Over the last several years, experimentation with Web-based applications was very high. It seemed that if you could spell "Internet" you could get VC money to develop a new Web-based application. But, as the dot-com bust illustrates, the market uncertainty was high because of the many failed ventures. As expected, some services such as information portals (CNN) are successful, but other ideas, such as online grocery shopping, have cost venture capitalists hundreds of millions of dollars, which illustrates the great experimentation and high market uncertainty.

---

[3] In this case the IETF is acting as the market and selecting the technology with the best market fit. One can also argue that the IETF can standardize technologies that have already been selected by the market (the original HTTP is such an example).

The next assumptions discuss the conditions under which the advantage of experimentation and choice is not enough to outweigh the inefficiencies of a distributed management structure. The case studies in Chapters 8 and 9 (email and voice) of network-based services show how there are different ways to provide each of these services, with different management structures. For these particular services, there are clear business and technical advantages to the more centralized architecture.

> **ASSUMPTION 6**   For some services, Business and Technical Advantages (BTA) lead service providers to provide services that are more centrally managed. For these services, if MU is zero, then a centralized management structure makes sense.

> **ASSUMPTION 7**   There are services for which market uncertainty is low relative to BTA, and this uncertainty will remain low with high confidence.

Some reasons for low market uncertainty are regulation that may prohibit choice (such as the early history of voice services), service providers that have learned from previous attempts to provide the service, or technology that is mature, such as PBXs in the mid and late 1980s. It is important to understand that this high confidence of low market uncertainty does not include paradigm shifts (for example, the SPC PBXs — see Chapter 9). Paradigm shifts are generally not predictable.

> **RULE 2**   If $(E[Max(X(1),..., X(n))] - E(X)) < BTA$, a service provider should consider providing a service with a more centrally managed architecture.

That is, if the advantage of market uncertainty combined with n simultaneous experiments is less than the business and technical advantages of central management, then a centralized management structure will work best. Furthermore, a single service choice with more efficient centralized management will meet market needs as well as several different choices with a less efficient distributed management structure because the low market uncertainty makes it easy to always meet the market.

This theory looks at only a single generation of a service. This is not realistic because services evolve over time. Later, I expand this theory by incorporating the evolution of services over time; in each generation, there will be n different attempts (experiments) to provide a service with a good market match. Thus, each service generation is composed of many service instances from simultaneous experimentation (that is, a service group), which are the efforts of one or many different contributors. This theory incorporates service providers' learning from previous generations of experiments, thus reducing the market uncertainty from generation to generation.

ASSUMPTION 8   As a service evolves over many successive generations, each generation of the service consists of a group of service experiments, with each experiment producing a service instance. Services exist for which the market uncertainty decreases in a predictable manner as a function of the service generation.

RULE 3 Service providers are likely to succeed at providing a service with centralized management in the first generation of the service when the advantage of MU and parallel experimentation do not outweigh BTA. The service provider should wait no longer to provide the more centralized architecture than the generation of the service when the advantage of MU and parallel experimentation summed over all future generations will never overcome BTA.

This rule is used to decide the upper and lower bound of when to switch management structures of a network-based service. When to migrate depends on the number of generations a service is expected to evolve. Longer evolution implies that experimentation will still be of more value than central management because over time the service will continue to come closer to meeting the changing market.

ASSUMPTION 9   Technology changes the range of possible services.

One example of a technology change that completely changed a market was the leap from a step-by-step PBX to SPC architecture, as discussed in Chapter 9.

RULE 4 If technology changes, market uncertainty may increase.

Another example of this is VoIP, as discussed in Chapter 10. This new paradigm to provide voice over data packet networks is creating large amounts of uncertainty about the future of voice services. It is very different from the current technology of switched circuits. For now, nobody is sure what features will have the most value to users as this new technology evolves.

This theory is fundamental to understanding how to design the management structure of network services. It provides a framework to analyze the management structure of a service with respect to market uncertainty and the number of experimental attempts to provide the service compared to the potential advantage of centrally managing the service. It shows that when a centrally managed service has advantages from a business and/or technical perspective, the market for the service may still be better met with services that have a distributed management structure because such a management structure allows more experimentation with features. When users have more choices that are more diverse, the winning service provider is able to reap huge profits because its service offering is a superior market match.

The next section provides a framework for a mathematical model showing Theorem 1 expressed in terms of a classic model in statistics known as maximum or extreme order statistics [7].

The Value of Many Experiments — Extreme Order Statistics

This section explains how providing a user with choice creates value for the service providers and how this value increases as market uncertainty grows. As shown previously, it is difficult to match services to markets when market uncertainty is high. To a single service provider, providing a new and innovative service is a gamble: Sometimes you win, sometimes you lose, and sometimes it takes a while to determine if you've won or lost. Service providers cannot predict how well a service will match the market (see Assumption 1). Many service providers can experiment creating n service instances, then let the users pick the best outcome (as stated in Rule 1 above).[4] The expected outcome is much higher. Picking the best of many experiments has the potential to greatly exceed the expected value of a single experiment.

Assuming a normal distribution for the value of an experiment, Figure A.1 shows what we expect to happen by attempting several parallel experiments for a particular service. It shows the probability of experiments being a particular distance from the mean. $V = E(X)$ denotes the expected value of a particular experiment. Looking at the percentages in Figure A.1, we expect that 34 percent of the experiments will fall between the mean and +1 standard deviation from it, 13.5 percent between 1 and 2 standard deviations, and 2 percent between 2 and 3 standard deviations from the mean. This matches the simulation results in Figure 6.1 from Chapter 6. To find a superior service we expect to need more than 769 experiments to find one that has a value greater than +3 standard deviations from the mean. This illustrates that finding great services may take on the order of 1,000 attempts.

---

[4] We are assuming that the parallel experiments have no correlation to simplify the mathematics, but the results still hold no matter what the correlation.

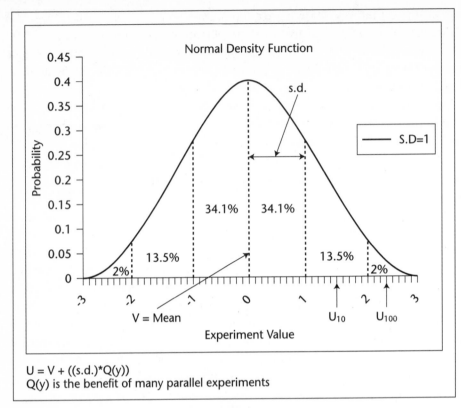

Normal Density Function

Figure A.1 Best of many experiments (value of many experiments).

Figure A.1 shows $U(10)$ and $U(100)$, the expected maximum of 10/100 experiments. That is, $U(10)$ is the value of the best experiment from a sample of 10 experiments. This maximum is composed of two different components: first, the effect of the mean and, second, the offset from the mean. This offset from the mean (V) is itself composed of two parts: first, the effect of the standard deviation and, second, the effect of the parallel experimentation. Thus, I can express $U(n)$ in terms of these parts: $U(n) = V + Q(n)*S.D$. That is, the maximum of n experiments equals the distribution mean plus the value of n experiments times the standard deviation of the normal distribution. $Q(n)$ measures how many standard deviations from the mean $U(n)$ is. Intuitively it makes sense that $U(n) >= V$ because to get an expected value for the mean, we do the n experiments, take the first one (expected value = V), and disregard the rest. It follows that the probability of $U(n)$ greatly exceeding V increases as n or the variance grows.

Roughly, for $n = 2$, $Q(n) = .85$, for $n = 10$, $Q(n) = 1.5$, for $n = 100$, $Q(n) = 2.5$, and for $n = 1,000$, $Q(n) = 3$, again matching the results in Figure 6.1. The intuition behind this is that, as you increase the number of experiments, the best of these experiments has a value that grows farther from the mean, but at a decreasing rate. For example, with 10 experiments you expect one of the outcomes to be between one and two standard deviations from the mean, but an outcome greater than three standard deviations from the mean is likely to require 1,000 experiments.

As uncertainty increases, so does the gain from experimentation and thus the potential for profit. To see how this works, consider the following example: Let the $S.D. = 1$ and $n = 10$ with a mean of zero. With $n = 10$, $Q(10) = 1.5$, so $U = 1 * 1.5 = 1.5$. If we increase the standard deviation to 2, then $U = 2 * 1.5 = 3$. This example shows that $Q(n)$ is a measure of how many standard deviations U is away from the mean.

This model, based on the best of many service experiments, is options-based because many service providers create several options for a particular service that users can pick from. When only a single choice for a service exists, the expected value is lower than if the user has many choices. The model illustrates how uncertainty increases the benefit of many choices.

The methods of order statistics allow us to understand the benefit of many parallel experiments relative to a single experimental attempt at providing an instance of a service. I assume a normal distribution and experiments that are not correlated, but the basic idea holds for any distribution or correlation between experiments. The expected value of the best of many experiments may greatly exceed the mean and is always at least as good as the expected value of a single experiment. The next section uses these results to model the expected value a service provider receives by providing the best service, which users have selected from many choices.

Mathematical Model

This appendix quantifies the theory by presenting one possible mathematical model based on it and the extreme order statistics discussed previously. This model is similar in nature to the options-based approach by Baldwin and Clark [1], which explains the value of modularity in computer systems design and follows previous papers by Bradner and myself about the advantages of modular design in standards [2][3][4]. This model has two stages. First, from Rule 2, the model views services at a particular generation; next, Rule 3 expands my model to study how services evolve over many generations. At each evolutionary stage of a service, I model learning from the previous generation with a learning function. This multigeneration model allows one to predict at what generation of the service a more centralized management structure may start to make economic sense.

This model focuses on two main forces affecting the value that providers receive for their services. First is the benefit of many parallel experiments combined with market uncertainty that pushes services to a more distributed management structure; next is the efficiency of centralized management that pulls services to centralized architectures. The model is based on the premises that environments providing easy experimentation may not provide the optimal management structure and that environments optimized for efficient service management may not be conducive to numerous experiments.

Modeling a Single Generation of a Service

As before, MU represents the market uncertainty, as discussed in Assumption 1; it is the random value to the service provider of providing a particular instance of a service. As Figure A.1 shows, V is the expected value of X, that is, $E(X) = V$. By the definition of standard deviation (S.D.), $S.D.(X) = MU$; that is, the standard deviation of the random variable denoting the value of a service to its provider is equal to the market uncertainty. MU is defined as the inability to predict the value service providers receive for a particular service instance, which is just a measure of the variability of the distribution of X.

In this model, the Business and Technical Advantage (BTA) of a centrally managed service, relative to a more distributed management style, is represented as a cost difference. BTA is the total advantage achieved by offering the centrally managed service. It may include both management and technical components. BTA is very general, as it must capture all the advantages of centralized management.

Let $CP(L)$ be the cost to provide services with management structure L. E is for end-2-end type services (distributed management), C is for centralized management structure. This cost is comprehensive and includes both the internal and external components, including internal infrastructure, equipment (including software), and management.

Using this terminology, Assumption 6 can be restated as $CP(E) > CP(C)$. It is more expensive to provide services with distributed management than with centralized management. Thus, the equation for BTA is:

Equation 1: $BTA = CP(E) - CP(C)$

$VP(L)$ is the expected value to a service provider with a particular management structure L. This value is the total value the provider receives for providing the service minus the total cost of providing the service. For a service with a central management structure that allows only one service instance, the value is:

Equation 2: $VP(C) = V - CP(C)$

For end-based services, I assume n service instances in a service group and allow market selection to pick the best outcome as defined in Rule 1. Remember that $Q(n)$ denotes the value of parallel experimentation; thus, the value of the best service at the edge with the benefit of experimentation in uncertain markets factored in is:

Equation 3: $VP(E) = V - CP(E) + MU*Q(n)$

The main difference between Equation 2 and 3 is the additional term representing the marginal value of experimentation. When there are n experimental attempts at a service and users are allowed to pick the best one of them, then this additional value is $MU*Q(n)$. This extra value depends on both the market uncertainty (MU) and the number of standard deviations away from the mean the best service will be (as discussed in this chapter and in Chapters 6 and 7).

Distributed management is better if **VP(E) - VP(C) > 0 => MU*Q(n) > CP(E) - CP(C)**, which is equivalent to **MU*Q(n) > BTA**. It is better to provide the service with a distributed management structure if:

Equation 4: $MU*Q(n) > BTA$

This equation states that the value of experimentation is greater than the business and technical advantages of centralized management. This indicates that the central management structure will be too confining to allow the experimentation required to meet the uncertain market. When equation 4 is true, the benefit of giving users choices is too great to ignore. This is the situation where the only way to figure out what users want is by giving them many different choices and seeing what they prefer.

This shows Rule 2: As market uncertainty increases, end-based services become more attractive due to the enhanced value of experimentation. If the cost differential between providing services with central management versus distributed management is less than the benefit gained from high market uncertainty and parallel experimentation, then the best service with distributed management has greater expected value than a single attempt to provide the service within the network.

This basic model is applied on two very simple cases: First, the case with only a single experiment and, second, when market uncertainty is zero. In both these cases, the advantage of environments allowing parallel experimentation is zero. Following these simple cases is a discussion of the more general case where experimentation is possible, and nonzero market uncertainty may make the effort of experimentation worthwhile.

These cases are simple to analyze. With only a single experiment there is no extra benefit to architectures that allow easy experimentation. **Q(n)** as defined becomes $0.^5$ "No uncertainty" means hitting the market every time; having more experiments is of no value because all experiments satisfy the market perfectly. In such cases, using the most efficient architecture makes sense because experimentation does not help.

A state of no market uncertainty is common with mature technology, legacy technology, or when legal requirements dictate services. Several examples of regulations modifying uncertainty for services are these: requirement of next-generation cell phones to support 911 location tracking, 911 location tracking behind PBXs, and hearing aid compatibility of phones in public locations or workplaces.

The preceding cases are not interesting; the service provider does the obvious by providing the service in the most efficient way. Following is the more interesting case where the best management structure is not clear. On one hand, the advantages of market uncertainty and parallel experimentation tend to favor environments that promote experimentation; on the other hand, efficiencies of centralized management of services may outweigh these advantages.

Assume that market uncertainty exists and the environment allows parallel experimentation, as discussed in Assumptions 1 and 2. Figure A.2(a) shows the relationship between MU (market uncertainty), BTA (business and technical advantage transformed into a cost differential), and n, the number of experiments run in parallel. This surface shows the relationship for a range of n (the number of simultaneous service experiments) between 1 and 20. Points on the surface show where market uncertainty equals $BTA/Q(n)$; the points above the surface show where services work well with end-2-end architecture because of the advantage of parallel experiments and market uncertainty. Points below the surface have low enough market uncertainty relative to BTA that centralized architectures should meet market needs. The forward edge of the surface shows the amount of MU required to offset BTA for a single experiment.[6]

From here, the surface slopes sharply down with regard to the number of experiments, showing the great value of experimentation. This is as expected because the range of services benefiting from end-2-end architectures grows with more experimentation. In addition, as expected, this growth is at a decreasing rate. The rate of decrease levels out quickly, at around 10 experiments, showing that the biggest gain from parallel experimentation is from relatively few experiments.

[5] It is more complex than this. The value is 0 if services with negative values must be kept, which may be the case with internal network-based services. However, keeping only positive outcomes, as end-2-end architecture tends to allow, raises this value to 0.39.

[6] This is greater than the mean, since we reject experiments without outcomes less than zero.

As expected from Rule 1, the value of a service to its provider increases at a slower rate with respect to n, the number of experiments. It is increasing at a constant rate with respect to MU, the market uncertainty. The surface in Figure A.2(b) illustrates this by showing the value (Z-axis) of running n (Y-axis) experiments with regard to the MU (X-axis). The curved lines for increasing n show the decreasing rate of increase, while the straight lines for increasing MU show the linear increase with regard to MU. This simple model for a single generation of a service fits the theory well.

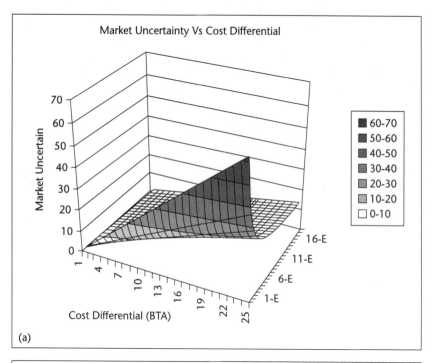

(a)

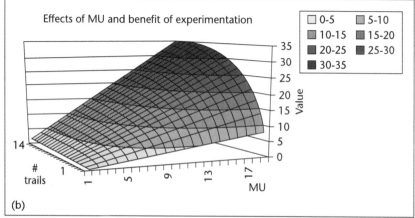

(b)

Figure A.2 Simple model of giving users choices.

This model provides a framework to help understand the relationship between market uncertainties, many parallel experiments, and the advantages of a centrally managed service. The surfaces allow visualization of these trade-offs that affect the choice of management structure. The model helps the manager and investor understand that uncertain markets can still be of high value with the correct strategy.

Below, I expand this basic model to illustrate how services change from generation to generation as they search for a better match in uncertain markets. This section introduces learning — that is, service providers gain experience from the previous generation about the preferences in the market.

Learning may or may not occur between service generations; if it does, the learning rate may be different at each generation. We do expect that the effect of learning should decrease from generation to generation as the technology matures and market preferences become more focused. Figure A.3 shows an example of how to represent learning for the normal distribution. The effect of learning is to squash the curve by decreasing the standard deviation (that is, market uncertainty). Learning has the effect of reducing the benefit of many experiments because each experiment falls within an increasingly narrow range centered on the mean, as Figure 6.1 shows; thus, the value of many experiments decreases.

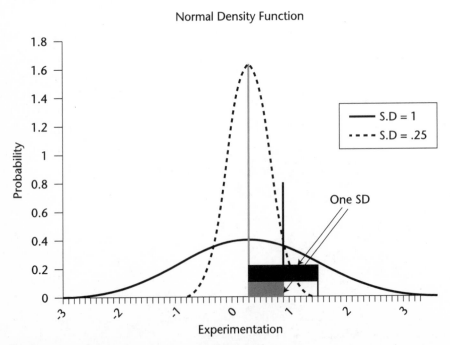

Figure A.3 Learning with a normal distribution.

This model views services as evolving over multiple generations, where each generation learns from the past. Figure A.3 shows this as a continued compression of the distribution. Using difference equations, I define the value of a service at the nth generation based on its value from the previous $(n-1^{th}$ generation) plus the additional value gained in the nth generation. A function dependent on the generation models the effect of learning by decreasing market uncertainty at each generation. Let f(generation) be this learning function that decreases learning by the correct amount at the ith generation:

$f(0) = 1$ by definition because there is no decrease in **MU** at the first generation.

Equation 5: $f(x) \in [0,1], x = 1,2,...,$

I assume this learning is symmetric; that is, all service providers learn the same for all experiments run by everybody.[7]

Derived from the preceding single-generation model (Equations 2 and 3), the following equations represent the value of the first generation of this multigeneration model:

Equation 6: $VS_1(C) = V_1 - CP(C)$
Equation 7: $VS_1(E) = V_1 - CP(E) + MU*Q(n)$
The value of the nth generation is 8:
Equation 8: $VS_n(C) = VS_{n-1}(C) + V_n$
Equation 9: $VS_n(E) = VS_{n-1}(E) + V_n + f(n) * Mu_n * Q(y_n)$

The value of the first generation of a service is identical to the previous single-generation model. The value of the nth generation of a centrally managed service is the value of the service at generation $n - 1$, plus the new value gained from the current (n^{th}) generation, assuming a single attempt to provide the service. Likewise, the value of a network-based service with a distributed management structure is the same for the first generation as shown previously.[8] For the nth generation, the value of the service is the value at the $n - 1^{th}$ generation, plus the additional benefit of picking the best experiment from the n attempts at this new generation (with learning factored in). Solving these difference equations gives:

Equation 10: $VP_n(E) = \sum_{i=1}^{n} V_i - CP(E)_n + MU * Q(y) \sum_{i=1}^{n} f(i)$

Equation 11: $VP_n(C) = \sum_{i=1}^{n} V_i - CP(C)_n$

This illustrates that the greater value of providing a service with distributed management is the sum of advantages gained from all previous generations. Thus, the benefit is dependent on the sum of $f(i)$ over all

---

[7] We know this is not true — some people just don't learn — but this assumption makes the problem tractable.

[8] To make a simple model we are folding in the cost from the previous generations.

generations (to infinity for the limit), times the gain from experimentation with market uncertainty. The major factor affecting this sum is the rate of decrease of $f(i)$, the learning rate. Figure A.4 shows several different types of learning functions, from the base case with no decrease, to a slowly decreasing harmonic series, and, finally, in a rapidly converging geometric progression. This decrease in learning fits into two different groups: functions that sum to infinite and functions that converge to a limit, as n approaches infinity. Figure A.4 shows both types: First, no decrease or the harmonic decrease (specifically, market uncertainty at the ith generation is reduced by $1/i$) which sums to infinity; second, two converging geometric decreases (specifically, market uncertainty at the ith generation is reduced by ai, $a < 0$), for $a = 1/2$ and $1/4$.

Different types of learning functions invoke dramatically different behavior. A learning function that diverges implies that a long evolution overcomes any advantage of a more centrally managed service. For example, Equation 12 shows what happens with the divergent harmonic series because it sums to infinity; if the service provider is willing to keep evolving the service, any cost advantage will be overcome because experimentation keeps adding value.

$$\text{Equation 12: } VP_n(E) = \sum_{i=1}^{n} V_i - CP(E) + MU * Q(y) \sum_{i=1}^{n} \frac{l}{i}$$

A convergent learning rate, such as any geometric progression, strictly limits the advantage gained from market uncertainty and many experiments. Below is an example of the convergent geometric series (specifically, it converges to $1/(1 - a)$). In this case a service provider never overcomes more than **MU*Q(n)*(1/(1 - a))**, even if the service evolves forever.

$$\text{Equation 13: } VP_n(E) = \sum_{i=1}^{n} V_i - CP(E) + MU * Q(y) \sum_{i=1}^{n} a^i, (a < 1)$$

The preceding equations allow one to compute the value of services at any generation, even an infinite amount. This allows a similar analysis to that shown in Figure A.2, but with a fixed number of experiments (that is, 10). In the next four surfaces, the Y-axis becomes the generation number, not the number of experiments as in Figure A.2. Figure A.5 and Figure A.6

show examples of the resulting surface for different learning curves (the last y value is for $n = $ infinity, showing the long-term effects of evolution). In Figure A.5(a), there is no learning ($f(i) = 1$, for all i), showing a fast decrease in the amount of **MU** required to overpower **BTA** as the service evolves. As the service is allowed to evolve for more generations, the amount of **MU** required to justify a distributed end-2-end management structure decreases. At each generation, the gain from experimentation is the same. Figure A.5(b) shows a decrease in market uncertainty by $1/n$ at the nth generation. Overcoming any cost advantage (**BTA**) of centralized services is still possible, as this figure shows, but it happens more slowly than with no learning. These surfaces show a very different situation than in Figure A.6, where a convergent geometric series (as in Equation 13) represents the learning function. In both these figures, the limit to which the series converges bounds the **BTA** that experimentation will overcome. Figure A.6(b) has a series that converges faster than (a), which illustrates the limited value of experimentation because in later generations market uncertainty is very low.

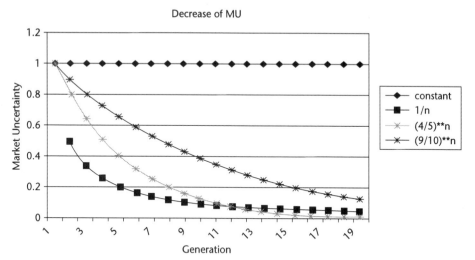

Figure A.4   Different learning rates.

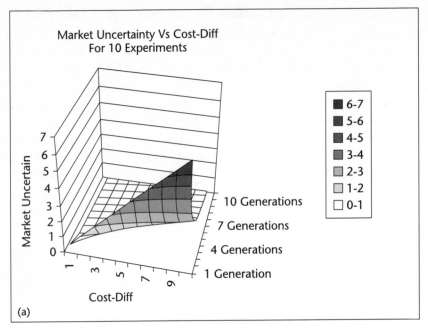

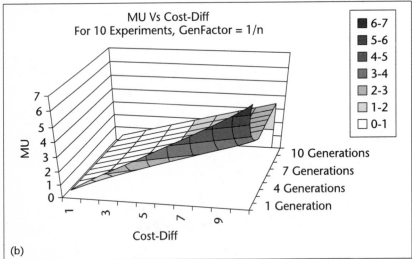

Figure A.5   No learning compared to harmonic learning ($1/n$).

The previous graphs give a framework for examining the trade-offs among market uncertainty, any advantages to a centrally managed service, and the number of generations the service is expected to evolve, given a fixed number of experiments. One important question is when, if ever, will the benefits of a centrally managed service overcome the advantage of experimentation to meet the market. It needs to be determined at what generation in the evolution of a service the advantages of experimentation is small compared to the efficiencies of a centralized management structure, as discussed in Rule 3. From the previous equations, it is possible to find both lower and upper boundaries for when a centralized management structure is likely to be successful in meeting market demands.

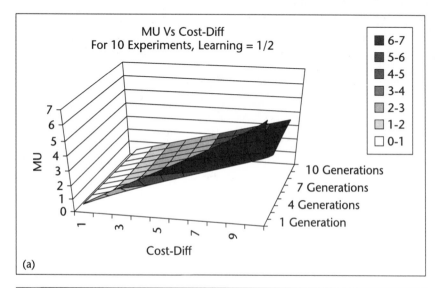

(a)

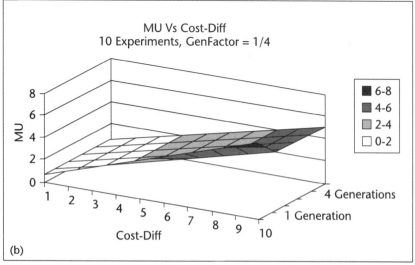

(b)

Figure A.6 Geometric learning $(1/i)^n$.

The lower boundary is the generation at which a provider should first consider providing a more centrally managed service. This is the generation of the service when the advantage to a centralized service structure is greater than the advantage gained from experimentation. This is the first generation when a central management structure has advantage. That is, I expect a centralized service to succeed at the ith generation when $MU*Q(n)*f(i) < BTA$. This is a lower boundary because if the service is in its last generation, then it pays to manage it centrally; however, continuing evolution of the service can still overcome greater BTA. The upper boundary is the longest amount of time to wait before shifting resources to services with a centralized management structure. It is the ith generation when the cost advantage of a centrally managed service can never be overcome with more experimentation. This is the generation when the advantage of central management will never be overcome by the benefits of experimentation. This is true when:

Equation 14: $$Q(y)*MU \sum_{i=n+1}^{\infty} f(i) < BTA$$

That is, it is true when the business and technical advantage of managing the service centrally is greater than the sum of benefits from experimentation from the current generation and future generations. This forms a bounded region when used in conjunction with the lower boundary. This shows the point at which one should consider a centralized management structure when designing network-based services.

Figure A.7(a) illustrates this lower bound for several different learning functions, including one example of a divergent series (specifically, harmonic) and several different examples of geometric series that converge at different rates to a finite sum. It shows that the harmonic series initially may decrease market uncertainty faster, but in the end, any geometric series will decrease learning at a much faster rate because of its convergence. Figure A.7(b) shows this upper bound. As expected, there is no upper bound for any divergent series (specifically, harmonic) because any cost advantage of a more centralized managed service can be overcome as long as the service provider is willing to continue evolving the service forever.

One important question is whether it is better to have fewer generations of a service with more experimentation per generation or more generations of the service with less experimentation per generation. With constant **MU** (that is, no learning between generations), the slowing rate of increase of **Q(n)** implies that more generations with less experimentation are best. If **MU** does decrease, it limits the gain from experimentation, making the answer dependent on the rate of decrease. This is one area of future research.

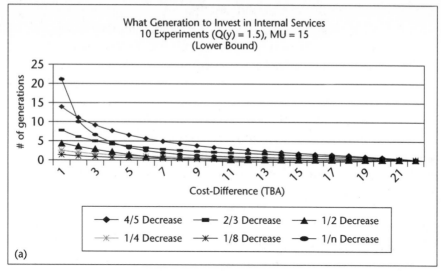

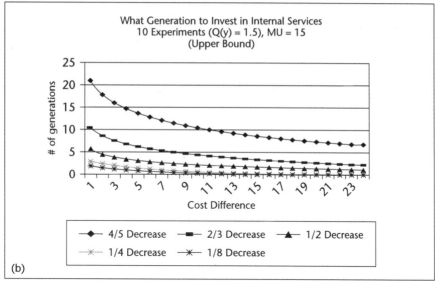

Figure A.7   Lower and upper bound of switching point to centralized management.

The previously discussed theory and model provide a framework to better understand the choices a service provider must make when deciding how to design a management structure when providing a service. When the market is not predictable by a service provider, this model helps to clarify the trade-offs among any possible advantage of a more centrally managed structure, the number of experiments all service providers undertake, and the number of evolutionary generations a service is expected to undergo. When you do not know what users want, single attempts to create services with centralized management schemes are unlikely to meet market demands. Instead, allowing the market to select the best service from many parallel experiments will be more successful at meeting the market. When user understanding of a technology has sufficiently evolved, then the end-2-end (distributed) architecture that allows easy experimentation will not meet market needs any better than more efficient centrally managed services. The close clustering of all experiments makes it easy to satisfy the user. While experimentation with services helps meet uncertain markets by giving users a wide range of service offerings from which to choose, this benefit is greatest in the first 10 to 20 experiments. Finally, I demonstrate the value of a service as it evolves from generation to generation and the effect of learning from previous generations.

This model illustrates that the ideal management structure for a network-based service changes as market uncertainty decreases (or increases with new technology). The model captures the chaotic behavior that occurs when the environment is dynamic. It allows visualization of the trade-offs involved in deciding how to manage network-based services. Managers and investors who understand these trade-offs have a competitive advantage over those who don't because they can tailor the management structure to maximize value.

Applying the Model

To apply this model one must estimate the market uncertainty (**MU**), the business and technical advantage (**BTA**), and the rate at which learning reduces **MU**. These items are hard to measure precisely. While I used real numbers to produce the graphs that show the trade-offs involved, how to get these numbers is not known. These factors should be estimated in terms of being high, medium, or low. Some progress in measuring **MU** by McCormack [8][9] and Tushman [10] is discussed in Chapter 6. The measurement of **BTA** is a combination of technical advantage and business cost savings.

Many services, such as email (see Chapter 8), evolve in multiple generations. First, email was an intra-company service seldom used to communicate outside the organization. Standards-based systems, such as the text-based Internet, followed. Next, MIME allowed attachments to Internet email. Finally, we arrived at the centralized Web-based email systems that have become popular in the last five years. Similar to **MU** and **BTA**, estimates of the rate of decrease per generation of **MU** are hard to quantify, allowing only coarse-grained estimates at this time; the most important attribute is the divergent or convergent nature of the learning. As this theory shows, the way **MU** decreases may limit the benefit of parallel experiments.

One way to view the relative success of flexible decentralized services compared to efficient centralized services is the percent of the market captured by each group. This is what Figure A.8 illustrates. Features incubate in the outer region where the market selects the successful features for inner migration. Selected services move from the outer edge inward, toward the center; the closer a service is to the center, the more centralized it is. As new technologies come and go, I expect the inner core to grow and shrink according to the market uncertainty. If **MU** is increased because of new technology, then services will shift from the core to the outer edges. After learning reduces **MU,** the successful features migrate into the core. This figure captures the success of a particular management structure and its dynamic nature.

One example of a real situation mapped to Figure A.8 is the PBX versus Centrex market of the late 70s and early 80s. Before the PBX switched to SPC architecture in the mid 70s, it was a very stable technology. Centrex, the centralized version of PBX services, was a stable technology; both had pieces of the market based on attributes other than feature sets. With the introduction of the new SPC architecture, however, the feature set of PBXs exploded and market uncertainty increased. This caused the percentage of PBXs to grow at the expense of Centrex because this new generation of PBXs matched users' needs better; Figure A.8 illustrates this as a growing of the inner region with stable services.

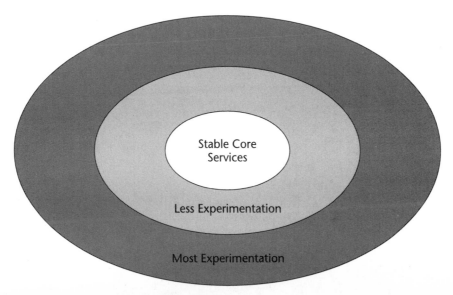

Figure A.8 Model management diagram.

This model shows the importance of having both regions in Figure A.8. On one hand, the ability to provide end-2-end services is necessary to meet user needs in uncertain markets. The ability to try out many different features for services when market uncertainty is high is the best way to understand the market. After understanding customer needs, the ability to migrate a feature into the network is necessary to capitalize on the business and technical advantages of centralized management. The outer region shows the power of innovation, while the inner region allows for efficient implementations of the best ideas.

Email and voice services are examples of services that have this two-layer structure. PBXs have provided a way to test and grow services that have become very successful in Centrex. Email maintains this ability to experiment with features (because of the underlying end-2-end architecture of Internet email) and also permits the centrally managed, core-based email services based on the same set of standards to adopt the successful features. Showing that these two different network-based services have a similar structure is strong evidence of the value of both regions in Figure A.8.

Conclusions

This theory and the model of the economic value of network services provide a framework to understand the advantages of experimentation and market uncertainty compared to the business and technology advantage of services with centralized management architectures. It shows that when users are confused (high market uncertainty), the value of experimentation is high. When service providers can predict what services and features will meet market demands, the management structure of the service becomes more important than the ability to innovate.

This work is one way to quantify the end-2-end argument, showing the value of end-2-end services due to their application independence of core network services. This work illustrates how end-2-end services will match markets best and produce the highest value to a service provider when high market uncertainty boosts the benefit of experimentation. It also shows that end-2-end architectures tend to lose some of their attraction as it becomes easier for service providers with more centralized structures to meet market needs as well, but at a lower cost and with other benefits.

This work helps link innovation in network design to the architecture of service management. It does so with a model based on a real options framework that illustrates market uncertainty, ease of experimentation, and the number of generations the service is expected to evolve. It is important to understand how and why successful services have evolved in the PSTN and Internet, which is especially significant in the age of convergence of data and voice services.

References

Chapter 1

[1]   Isenberg, David S. 1998. "The Dawn of the Stupid Network." *ACM Networker 2.1* (February/March): 24–31.

Part One

Chapter 2

[1]   Clark, Kim. 1985. "The interaction of design hierarchies and market concepts in technological evolution." *Research Policy* 14: 235–251.

[2]   Bradner, Scott. 2001. "Mapping a way forward." *Network World* (August 8): 30.

[3]   Ambeler, Chris. "Custom local area signaling services class" No. 111A ESS Div. 3 Sec. 1z3. hyperarchive.lcs.mit.edu/telecom-archives/archives/technical/class.ss7.features.

[4]   Lawrence, Paul and Jay Lorsch. 1986. *Organization and Environment*. Boston, MA: Harvard Business School Press.

Chapter 3

[1] Saltzer, Jerome, David Reed, and David Clark. 1984. "End-To-End Arguments in System Design." *ACM Transactions in Computer Systems* 2, 4 (November): 277–288.

[2] Branstad, D. K. "Security aspects of computer networks." AIAA paper No. 73-427, *AIAA Computer Network Systems Conference*, Huntsville, Alabama, April 1973.

[3] Diffie, W. and M. E. Hellman. 1976. "New directions in cryptography." *IEEE Transactions on Information Theory*, IT-22, 6 (November): 644–654.

[4] Hippel, E. 1998. "Economics of Product Development by Users: The Impact of 'Sticky' Local Information." *Management Science* 44(5).

[5] Berners-Lee, Tim. 1999. *Weaving the Web*. San Francisco: Harper.

[6] Reed, D. P. 1978. "Naming and synchronization in a decentralized computer system." Ph.D. thesis, MIT, Dept of EECS, September, Technical Report TR-205.

[7] Isenberg, David S. 1998. "The Dawn of the Stupid Network." *ACM Networker 2.1* (February/March): 24–31.

[8] Gaynor, M. 2001. "The effect of market uncertainty on the management structure of network-based services." Ph.D. thesis, Harvard University.

[9] Clark, David and Marjory S. Blumenthal. "Rethinking the design of the Internet: The end-to-end arguments vs. the brave new world." Submitted TRPC August 10, 2000 (http://www.ana.lcs.mit.edu/anaweb/PDF/Rethinking_2001.pdf).

[10] Carpenter, B. *RFC2775—Internet Transparency*. February 2000.

[11] Carpenter, B. *RFC1958—Architectural Principles of the Internet*. June 1996.

[12] Lawrence Lessig. 1999. *Code and Other Laws of Cyberspace*. N.Y.: Basic Books.

[13] Lawrence Lessig. 2002. *The Future of Ideas: The Fate of the Commons in a Connected World*. Vintage Books.

Chapter 4

[1] Laudon, Kenneth and Jane Laudon. 2002. *Management Information System*, 7th ed. NJ: Prentice Hall.

[2] Isenberg, David S. 1998. "The Dawn of the Stupid Network." *ACM Networker 2.1* (February/March): 24–31.

[3] Clark, Kim. 1985. "The interaction of design hierarchies and market concepts in technological evolution." *Research Policy* 14: 235–251.

[4] Shapiro, Carl and Hal Varian. 1998. *Information Rules*. Boston, MA: Harvard Business School Press.

[5] Porter, Michael. 1998. *Competitive Strategy*. NY: Free Press.

Chapter 5

[1] Merton, R. 1992. *Continuous-Time Finance*. Oxford, UK: Blackwell Publishing.

[2] Brealey, Richard and Stewart Myers. 1996. *Principles of Corporate Finance*. NY: McGraw-Hill.

[3] Amram, M. and N. Kulatilaka. 1999. *Real Options, Managing Strategic Investment in an Uncertain World*. Boston, MA: Harvard Business School Press.

[4] P. Balasubramanian, Nalin Kulatilaka, and John Storck. 1999. "Managing Information Technology Investments Using a Real-Options Approach." *Journal of Strategic Information Systems*, 9:1.

[5] Baldwin, Carliss and Kim Clark. 1999. *Design Rules: The Power of Modularity*. Cambridge, MA: MIT Press.

[6] Gaynor, M. and S. Bradner. 2001. Using Real Options to Value Modularity in Standards. *Knowledge Technology and Policy*. Special issue on IT Standardization. 14:2.

[7] Gaynor, M., S. Bradner, M. Iansiti, and H. T. Kung, HT. "The Real Options Approach to Standards for Building Network-based Services." *Proc. 2nd IEEE conference on Standardization and Innovation*. Boulder, CO, October 3, 2001.

Chapter 6

[1] Clark, Kim. 1985. "The interaction of design hierarchies and market concepts in technological evolution." *Research Policy* 14: 235–251.

[2] Berners-Lee, Tim. 1999. *Weaving the Web*. San Francisco: Harper.

[3] Tushman, Michael and Philip Anderson. 1986. "Technological Discontinuities and Organizational Environments." *Administrative Science Quarterly* 31: 439–465.

[4] MacCormack, Alan. 2000. "Towards a Contingent Model of the New Product Development Process: A Comparative Empirical Study." Working Paper 00-77, Harvard Business School, Division of Research.

[5] MacCormack, Alan and Roberto Verganti. 2001. "Managing the Sources of Uncertainty: Matching Process and Context in New Product Development." Working Paper 00-78, Harvard Business School, Division of Research.

[6] Hiller, Frederick S. and Gerald J. Lieberman. 1967. *Operations Research*: San Francisco: Holden-Day, p 631.

[7] Baldwin, Carliss and Kim Clark. 1999. *Design Rules: The Power of Modularity*. Cambridge, MA: MIT Press.

[8] Kauffman, S. 1993. *At Home In The Universe: The Search for Laws of Self Organization and Complexity*. Oxford, UK: University Press.

Chapter 7

[1] Lawrence, Paul and Jay Lorsch. 1986. *Organization and Environment*. Boston, MA: Harvard Business School Press.

[2] Clark, Kim. 1985. "The interaction of design hierarchies and market concepts in technological evolution." *Research Policy* 14: 235–251.

[3] Baldwin, Carliss and Kim Clark. 1999. *Design Rules: The Power of Modularity*. Cambridge, MA: MIT Press.

[4] Gaynor, M. and S. Bradner. 2001. "Using Real Options to Value Modularity in Standards." *Knowledge Technology and Policy*. Special issue on IT Standardization. 14:2.

[5] Gaynor, M., S. Bradner, M. Iansiti, and H. T. Kung. "The Real Options Approach to Standards for Building Network-based Services." *Proc. 2nd IEEE conference on Standardization and Innovation*. Boulder, CO, October 3, 2001.

Part Two

Chapter 8

[1] Gaynor, M. 2001. "The effect of market uncertainty on the management structure of network-based services." Ph.D. thesis, Harvard University.

[2] Gaynor, M. and S. Bradner. 2001. "Using Real Options to Value Modularity in Standards." *Knowledge Technology and Policy*. Special issue on IT Standardization. 14:2.

[3] Staff Writer. *What is the history of e-mail?* www.aptosjr.santacruz,k12 .ca.us/computers/page9.htm.

[4] Arnum, Eric. 2000. Year-end 1999 Mailbox Report, Messagingonline. www.messagingonline.com.

[5] Bhushan, Abhay, Ken Pogran, Ray Tomlinson, and Jim White. 1973. *RFC561—Standardizing Network Mail Headers*. sunsite.doc.ic.ac.uk/rfc /rfc561.txt, September 5.

[6] Postel, Jonathan. 1981. *RFC788—Simple Mail Transfer Protocol*. http: //www.ietf.org/rfc/rfc0788.txt?number=788, ISI, November.

[7]   Reynolds, J. K. 1984. *RFC918—Post Office Protocol.* http://www.ietf
      .org/rfc/rfc0918.txt?number=918, ISI, October.

[8]   Crispin, M. 1988. *RFC1064—Interactive Mail Access Protocol—Version 2.*
      http://www.ietf.org/rfc/rfc1064.txt?number=1064, SUMEX-AIM, July.

[9]   Borenstein, N. and N. Freed. 1992. *RFC-1341, Multipurpose Internet
      Mail Extensions.* http://www.ietf.org/rfc/rfc1341.txt?number=1341,
      June.

[10] Arnum, Eric. 2001. Year-end 2000 Mailbox Report, Messagingonline.
      www.messagingonline.com.

[11] McQuillan, John. 1984. "Office Automation Strategies—Electronic
      Mail: What Next?" *Business Communications Review* (June): 40.

[12] McQuillan, John. 1988. "Doing Business Electronically." *Business Com-
      munications Review* (May–June).

[13] Arnum, Eric. 1991. "Inbound Routing for E-mail/Fax Gateways."
      Business Communications Review (July).

[14] Arnum, Eric. 1993. "Electronic Mail Broadens Its Horizons." *Business
      Communications Review* (July).

[15] Delhagen, Kate, Emily Green, and Nick Allen. 1996. "E-mail Explodes."
      The Forrester Report 3: 7 (November), www.forrester.com/ER/Research/
      Report/.

[16] Brown, Eric, John McCarthy, and Michael Mavretic. 1996. "E-mail
      Shootout." *The Forrester Report* 3:7 (November), www.forrester.
      com/ER/Research/Report/.

[17] Passmore, Davie. 1996. "The End of Proprietary Email." *Business Com-
      munications Review* (June).

[18] Cooperstien, David and David Goodtree. 1998. "E-mail Outsourcing."
      The Forrester Report (December), www.forrester.com/ER/research/Brief.

[19] Browning, J., J. Graff, L. Latham. 1999. "Messaging Servers' Magic
      Quadrants: Enterprise and ISP." Strategic Analysis Report, The Gart-
      ner Group (May).

[20] Reid, Robert H. 1997. *Architects of the Web: 1,000 Days That Built the
      Future of Business.* New York: John Wiley & Sons.

[21] Freed, N. and N. Borenstein. 1996. *RFC2045—Multipurpose Internet
      Mail Extensions (MIME) Part One: Format of Internet Message Bodies,*
      November.

[22] Freed, N. and N. Borenstein. *1996. RFC2046—Multipurpose Internet
      Mail Extensions (MIME) Part Two: Media Types,* November.

[23] Moore, K. 1996. *RFC2047—Multipurpose Internet Mail Extensions
      (MIME) Part Three: Message Header Extensions for Non-ASCII Text,*
      November.

[24] Freed, N, J. Klensin, and J. Postel. *1996, RFC2048—Multipurpose Internet
      Mail Extensions (MIME) Part Four: Registration Procedures,* November.

[25] Freed, N. and N. Borenstein. 1996. *RFC2049—Multipurpose Internet Mail Extenstions (MIME) Part Five: Conformance Criteria and Examples,* November.

[26] Pelline, Jeff. 1997. "Yahoo buys Four11 for free email." www.news. cnet.com/news/0-1005-200-322847.html, *C/net News.com,* October 8.

[27] Pelline, Jeff. 1998. "Microsoft buys Hotmail." http://news.cnet.com /news/0-1004-200-325262.html, *C/net News.com,* January 3.

Chapter 9

[1] Sternberg, Michael. 1982. "Competitive Strategy in the Market for PBXs." *Business Communication Review* (July–August): 10.

[2] Alexander, Donald L. *1996. Telecommunications Policy—Have Regulators Dialed the Wrong Number?* Westport, CT: Praeger.

[3] Goeller, Leo. 1981. "Trends in PBX Design." *Business Communication Review* (September–October): 10–18.

[4] Bergman, Mark, and Klinck Courtney. 1984. "Evolution of the Modern PBX." *Business Communication Review* (July–August): 22–26.

[5] Sulkin, Alan. 1986. "Big Business for Small Business Systems." *Business Communications Review* (September–October): 8–11.

[6] Staff. 1981. "Voice Message Systems." *Business Communications Review* (January–February): 37–40.

[7] Kirvan, Paul. 1985. "Centrex Positions Itself for Your Future." *Business Communications Review* (July–August): 2.

[8] Brock, Gerald, W. 1994. *Telecommunication Policy for the Information Age—From Monopoly to Competition.* Cambridge, MA: Harvard University Press.

[9] Fernazbum, Thomas. 1988. "Market Gets the Voice Mail Message." *Business Communications Review* (September–October): 70–73.

[10] Kropper, Steven and Charles Thiesen. 1989. "Strategic Issues for Telcos in Enhanced Voice Messaging Systems." *Business Communications Review* (November): 51–53.

[11] Goeller, Leo. 1982. *Voice Communications in Business,* Chapter 20, abc TeleTraining. (First appeared as "A requiem for step," *BCR,* May–June 1980.)

[12] Christensen, Clayton. 1997. *The Innovator's Dilemma: When New Technologies Cause Great Firms to Fail.* Boston, MA: Harvard Business School Press.

[13] Bhushan, Brij. 1984. "Evolving Voice/Data Switch Technology." *Business Communications Review* (May–June): 29.

[14] Ricca, Mark. 1988. "Can Anyone Make a Profit in the PBX Market?" *Business Communications Review* (September–October): 33.

[15] 1982. "New Product, The Telephone is Becoming a Management Workstation." *Business Communications Review* (November–December): 34.

[16] Sulkin, Alan, 1989. "Digital Trend Builds in Key System Market." *Business Communications Review* (July): 73.

[17] Keeler, David. 1987. "The Key Telephone System Market Adapts for the 1990's." *Business Communications Review* (November–December): 7.

[18] Mikolas, Mark. 1991. "Five Myths about the PBX Market." *Business Communications Review* (March): 101.

[19] Kuehn, Richard. 1991. "Consultant's Corner: Musings from the ICA Show Floor." *Business Communications Review* (August): 77.

[20] Goldstone, Jerry. 1992. "Memo from the Publisher: The How and Why of Bringing PBXs Back into the Limelight." *Business Communications Review* (January): 4.

[21] Sulkin, Alan. 1993. "Directions in PBX Evolution." *Business Communications Review* (April).

[22] Sulkin, Alan. 1993. "Today's PBX Prepares for Tomorrow's Challenges." *Business Communications Review* (May).

[23] Sulkin, Alan. 1993. "PBX Losses Shrink During 1992." *Business Communications Review* (January): 32.

[24] Finneran, Michael. 1994. "PBXs Prepare to Miss the Next Boat." *Business Communications Review* (May).

[25] Borton, Gregory, and Lawrence Lutton. 1991. "Do ACD Features Meet Strategic Goals?" *Business Communications Review* (March): 37.

[26] Sulkin, Alan. 1992. "KTS/Hybrid Products: Small Systems, Big Features." *Business Communications Review* (July).

[27] Sulkin, Alan. 1995. "ACDs: Standalone vs. the PBX Vendors." *Business Communications Review* (June).

[28] Knight, Fred. 1992. "Rockwell's Spectrum Call Center System." *Business Communications Review* (October): 60.

[29] Sulkin, Alan. 1993. "Building the ACD-LAN Connection." *Business Communications Review* (April).

[30] Rosenberg, Arthur. 1981. "The Great Convergence: The Telephone Network and Messaging Systems." *Business Communications Review* (September–October): 19.

[31] Rosenberg, Arthur and Van Doren. 1983. "Voice Messaging—Today and Tomorrow." *Business Communications Review* (January–February): 9.

[32] Staff Writer. 1981. "New Product Voice Mail is Being Integrated into PBXs." *Business Communications Review* (September–October): 37.

[33] Byfield, E. T. 1981. "Voice Mail can benefit the Small Office." *Business Communications Review* (September–October): 23.

[34] Fross, Alan. 1989. "Centrex Revival Rests on Enhanced Services." *Business Communications Review* (April): 40.

[35] Rosenberg, Arthur. 1988. "Selecting a Voice Messaging System." *Business Communications Review* (March–April): 79.

[36] Fermazin, Thomas. 1988. "If Voice Messaging is the Answer, What is the Question?" *Business Communications Review* (October): 79.

[37] Fermazin, Thomas. 1989. "Voice Messaging isn't Always the Answer, What is the Question?" *Business Communications Review* (August): 74.

[38] Sulkin, Alan. 1989. "Hidden Costs of PBX Options." *Business Communications Review* (June): 40.

[39] Sulkin, Alan. 1991. "PBX Capabilities for the Mid-1990s." *Business Communications Review* (September): 51.

[40] Goldstone, Jerry. 1996. "Memo from the Publisher: The How and Why of Bringing PBX's Back into the Limelight." *Business Communications Review* (March): 4.

[41] Kuehn, Richard. 1997. "The PBX is Dead? Long Live the PBX!" *Business Communications Review* (May): 74.

[42] MITEL Corporaton. 1996. "Migrating to voiceLAN." *Business Communications Review* (October): special section.

[43] Sulkin, Alan. 2000. "PBX Market Gets Ready to Shift Gears." *Business Communications Review* (January): 43.

[44] Knight, Fred. 1997. "Big-Time Change is on the Way." *Business Communications Review* (July): 40.

[45] Passmore, Dave. 1997. column, *Business Communications Review* (December): 19.

[46] Goeller, Leo. 1986. "Ten Common Misconceptions about Centrex." *Business Communications Review* (December): 2.

[47] Sulkin, Alan. 1992. "Centrex Providers Discover that Small can be Beautiful." *Business Communications Review* (April): 58.

[48] Cook, Lloyd. 1981. "Integrated Site Networking: An Alternative to the Large PBX or Centrex." *Business Communications Review* (January): 43.

[49] Gordon, James, Margaret Klenke, and Cathy Camp. 1989. "Centrex." *Business Communications Review* (October): 39.

[50] Deans, David. 1991. "Are the Telcos Up to the Centrex Marketing Challenge?" *Business Communications Review* (June 1991): 50.

[51] Gable, Robert. 1994. "IXC Management Tools for 800 services." *Business Communications Review* (August): 35.

[52] PBX versus Centrex. www.commserv.ucsb.edu/hpage/hot/tac/centxucb.htm.

[53] Kuehn, Richard. 1992. "Consultants Corner: How Much Should Centrex Cost?" *Business Communications Review* (November).

[54] Horrell, Edward, "Secondary PBX Market Solidifies," *Business Communications Review*. (July–Aug 1988).

[55] Sulkin, Alan. 1989. "The 1988 PBX Market." *Business Communications Review* (January): 81.

[56] Sulkin, Alan. 1992. "PBX Market Tumble Continues." *Business Communications Review* (January): 23.

[57] Staff Writer. 1988. "New PBX and CO-Based ACD from Northern Telecom." *Business* Communications Review (July–August).

[58] Goldstone, Jerry. 1992. "ACDs and PBXs: Blurring the Lines." *Business Communications Review* .

[59] Fross, Alan. 1988. "The ACD Market Comes of Age." *Business Communications Review* (November–December).

[60] Gaynor, M. 2001. "The effect of market uncertainty on the management structure of network based services." Ph.D. thesis, Harvard University.

[61] Sykes, Dustin. 1988. "Voice Messaging: Brisk Growth and Critical Issues." *Business Communications Review* (November–December): 29.

Part Three

Chapter 10

[1] Handley, Schulzrinne, Schooler, and Rosenberg. 2000. *RFC2543-SIP: Session Initiation Protocol.*

[2] Greene, N., M. Ramalho, and B. Rosen. 2000. *RFC2805—Media Gateway Control Protocol Architecture and Requirements.*

[3] C. Huitema, J. Cameron, P. Mouchtaris, and D. Smyk. 1999. "An architecture for internet telephony service for residential customers." IEEE Network 13: 50–57.

[4] Xiaotao, Wu and Henning Schulzrinne. "Where Should Services Reside in Internet Telephony systems?" IP Telecom Services Workshop, Atlanta, Georgia, September 2000.

[5] AT&T. 1987. "5ESS Switch, The Premier Solution: Feature Handbook." September.

[6] Russell, Travis. 1998. Signaling System #7, 2nd ed. New York: NY: McGraw-Hill.

[7] P. Blatherwick, and R. Bell, and P. Holland. Jan. 2001. *RFC3054—Megaco IP Phone Media Gateway Application Profile.*

Chapter 11

[1] Amram, M. and N. Kulatilaka. 1999. *Real Options, Managing Strategic Investment in an Uncertain World.* Boston, MA: Harvard Business School Press.

[2] Gaynor, M. 2001. "The effect of market uncertainty on the management structure of network based services." Ph.D. Thesis, Harvard University.

[3] Rappaport, Theodore. 2002. *Wireless Communications—Principles and Practice,* 2nd ed. NJ: Prentice Hall.

[4] Smith, Clint and Daniel Collins. 2002. *3G Wireless Networks.* NY: McGraw-Hill.

[5] Gast, Matthew. 2002. *802.11 Wireless Networks—the Definitive Guide.* CA: O'Reilly.

[6] Flickenger, Rob. 2002. *Building 802.11 Wireless Community Networks.* CA: O'Reilly.

[7] Gaynor, M. and S. Bradner. 2001. "Using Real Options to Value Modularity in Standards." *Knowledge Technology and Policy.* Special issue on IT Standardization. 14: 2

[8] Gaynor, M., S. Bradner, M. Iansiti, and H. T. Kung. "The Real Options Approach to Standards for Building Network-based Services." *Proc. 2nd IEEE conference on Standardization and Innovation,* Boulder, CO, October 3, 2001.

Chapter 12

[1] Birrell, Andrew and Bruce Nelson. 1984. "Implementing Remote Procedure Calls." *ACM Transactions on Computer Systems,* 2: 1 (February).

[2] IBM Web site. Web Services Conceptual Architecture. http://www-3.ibm.com/software/solutions/webservices/pdf/WSCA.pdf.

[3] http://www.omg.org/gettingstarted/corbafaq.htm#TotallyNew.

[4] Microsoft Web site. DCOM: A business Overview http://msdn.microsoft.com/library/default.asp?url=/library/en-us/dndcom/html/msdn_dcombiz.asp.

[5] Microsoft Web site. XML Web services. http://msdn.microsoft.com/library/default.asp?url=/nhp/Default.asp?contentid=28000442.

[6] http://www.execpc.com/~gopalan/misc/compare.html. A Detailed Comparison of CORBA, DCOM, and Java/RMI.

[7] http://java.sun.com/webservices/faq.html#websrv.

[8] http://www.itl.nist.gov/fipspubs/fip161-2.htm. ELECTRONIC DATA INTERCHANGE (EDI).

[9]   Short, Scott. 2002. *Building XML Web Services for the Microsoft .NET Platform*. Redmond, WA: Microsoft Press.

[10] Graham, Steven, Simeon Simeonov, Toufic Boubez, Doug Davis, Glen Daniels, Yuichi Nakamura, and Ryo Neyama. 2002. *Building Web Services with Java, Making Sense of XML, SOAP, WSDL, and UDDI*. IN: Sams Publishing.

[11] Baldwin, Carliss, and Kim Clark. *Design Rules: The Power of Modularity*. Cambridge, MA: MIT Press. 1999.

[12] W3c Web site. http://www.w3.org/XML/.

[13] W3c Web site. http://www.w3.org/TR/SOAP/.

Appendix

[1]   Baldwin, Carliss, and Kim Clark. *Design Rules: The Power of Modularity*, Cambridge, MA: MIT Press. 1999.

[2]   Gaynor, Mark and Scott Bradner. 2001. "The Real Options Approach to Standardization." *Proceedings of Hawaii International Conference on Systems Science*.

[3]   Gaynor, M. and S. Bradner. 2001. "Using Real Options to Value Modularity in Standards." *Knowledge Technology and Policy*. Special issue on IT Standardization. 14: 2.

[4]   Gaynor, M., S. Bradner, M. Iansiti, and H. T. Kung. "The Real Options Approach to Standards for Building Network-based Services." *Proc. 2nd IEEE conference on Standardization and Innovation*. Boulder, CO, October 3, 2001.

[5]   Clark, Kim. 1985. "The interaction of design hierarchies and market concepts in technological evolution." *Research Policy*, 14: 235–251.

[6]   Amram, Martha and Nalin Kulatilaka. 1999. *Real Options, Managing Strategic Investment in an Uncertain World*. Boston, MA: Harvard Business Press.

[7]   Lindgren, B.W. 1968. *Statistical Theory*. New York: Macmillan.

[8]   MacCormack, Alan. 2000. "Towards a Contingent Model of the New Product Development Process: A Comparative Empirical Study." Working Paper 00-77, Harvard Business School, Division of Research.

[9]   MacCormack, Alan and Roberto Verganti. 2001. "Managing the Sources of Uncertainty: Matching Process and Context in New Product Development." Working Paper 00-78, Harvard Business School, Division of Research.

[10] Tushman, Michael and Philip Anderson. 1986. "Technological Discontinuities and Organizational Environments." *Administrative Science Quarterly* 31: 439–465.

Index

SYMBOLS AND NUMERICS

*## (star, number, number) telephone services, 21

2G cellular
Code Division Multiple Access (CDMA), 192
Global System for Mobile Communications (GSM), 193
IS-95 popularity, 193
Time Division Multiple Access (TDMA), 192

3G cellular
advantages/disadvantages, 207–208
bandwidth use efficiency, 193
development history, 192–194
future applications, 201–202
link layer protocols, 248
management structure, 194–196
packet-switched voice services, 193
spatial reuse efficiency, 208–210
voice capacity increases, 193
walled garden metaphor, 190–192

802.11 wireless networks
advantages/disadvantages, 205–207
Bay Area Wireless, 199

Boingo's business model, 199–200
centralized management, 198–199
community groups, 199
development history, 196–197
direct sequence spread spectrum (DSSS), 196
distributed management, 197–198
experimentation effects, 202–205
external antennas, 196
free public spectrum, 192
frequency hopping spread spectrum (FHSS), 196
future applications, 201–202
groundswell of activity, 191–192
IEEE standards, 196
Joltage business model, 199–200
link layer protocols, 248
management structure, 197–201
NoCatNet, 199
open garden metaphor, 190–192
spatial reuse efficiency, 208–210
uses, 191–192
wide-area applications, 196

A

ACD. *See* Automatic Call Distribution

ACR. *See* Automatic Call Routing
ADSI. *See* Analog Display Services Interface
airports, 802.11 wireless network support, 191
alternate routing, Centrex development history, 152
America Online (AOL)
email services, 55–57
ISP-based email, 124
Analog Display Services Interface (ADSI), caller ID, 51
application layer protocols, 248
arguments, conclusions, 245–247
ASCII, email development history, 118
assumptions
formal theory, 259–264
management theory, 98–101
Asynchronous Transfer Mode (ATM), high market uncertainty, 90
ATM networks, centralized management, 27
Automatic Call Distribution (ACD)
development history, 140–141, 145–147
network-based service, 16
Automatic Call Routing (ACR), PBX service, 62